Middle Level SSAT®
1000+ Practice Questions

Middle Level SSAT®: 1000+ Practice Questions
February 2024

Published in the United States of America by:

The Tutorverse, LLC

222 Broadway, 19th Floor

New York, NY 10038

Web: www.thetutorverse.com

Email: hello@thetutorverse.com

Copyright © 2024 The Tutorverse, LLC. All rights reserved. Except as permitted under the Copyright Act of 1976, no part of this publication may be reproduced or distributed in any forms or by any means, or stored in a database or retrieval system, without the prior written permission of the publisher.

Third party materials used to supplement this book may be subject to copyright protection vested in their respective publishers and are likewise not reproducible under any circumstances.

For information about buying this title in bulk or to place a special order, please contact us at hello@thetutorverse.com.

ISBN-13: 978-1-7321677-2-8
ISBN-10: 1-7321677-2-9

SSAT® is a registered trademark of the Secondary School Admission Test Board, Inc., which was not involved in the production of, and does not endorse, sponsor, or certify this product.

Neither the author or publisher claim any responsibility for the accuracy and appropriateness of the content in this book, nor do they claim any responsibility over the outcome of students who use these materials.

The views and opinions expressed in this book do not necessarily reflect the official policy, position, or point of view of the author or publisher. Such views and opinions do not constitute an endorsement to perform, attempt to perform, or otherwise emulate any procedures, experiments, etc. described in any of the passages, excerpts, adaptations, cited materials, or similar information. Such information is included only to facilitate the development of questions, answer choices, and answer explanations for purposes of preparing for the SSAT®.

**Available in Print & eBook Formats
visit thetutorverse.com/books**

**View or Download Answer Explanations
at thetutorverse.com/books**

Table of Contents

Welcome _____ 7

How to Use This Book _____ 8

Diagnostic Practice Test (Form A) _____ 11

Quantitative _____ 43

    Number Concepts & Operations _____ 45
        Place Value _____ 45
        Decimals _____ 46
        Fractions _____ 47
        Percents _____ 48
        Decimals/Fractions/Percents _____ 50
        Whole Numbers _____ 52
        Order of Operations _____ 53

    Pre-Algebra _____ 54
        Ratio and Proportions _____ 54
        Sequences, Patterns and Logic _____ 57
        Estimation _____ 58

    Algebra _____ 60
        Interpreting Variables _____ 60
        Solving Equations and Inequalities _____ 62
        Multi-Step Word Problems _____ 64

    Geometry _____ 67
        Pythagorean Theorem _____ 67
        Coordinate Plane _____ 68
        Transformations _____ 71
        Circles _____ 74
        Two and Three Dimensional Shapes _____ 76
        Spatial Reasoning _____ 77

    Measurement _____ 78
        Time and Money _____ 78
        Area, Perimeter and Volume _____ 80
        Angles _____ 82
        Unit Analysis _____ 84

    Data Analysis _____ 85
        Interpreting Bar Graphs _____ 85
        Interpreting Histograms _____ 88
        Interpreting Line Graphs _____ 91
        Interpreting Circle Graphs _____ 93

    Statistics & Probability _____ 95
        Basic Probability _____ 95
        Compound Events _____ 98
        Mean, Median, Mode _____ 99

Verbal *102*

Synonyms **104**
- Introductory 105
- Intermediate 106
- Advanced 108

Analogies **110**
- Guided Practice 111
- Mixed Practice 119

Reading *120*

Fiction 122

Nonfiction 132

The Writing Sample *142*

Final Practice Test (Form B) *144*

Answer Keys *177*

Diagnostic Practice Test (Form A) Answer Key **177**

Quantitative **178**

Verbal – Synonyms **180**

Verbal – Analogies **180**

Reading Comprehension **181**

Final Practice Test (Form B) Answer Key **182**

Middle Level SSAT® 1000+ Practice Questions

Welcome

Dear Students, Parents, and Educators,

Welcome to The Tutorverse!

You've just taken a positive first step toward acing the Middle Level SSAT - congratulations! This workbook contains the key to scoring well on the test: high-quality, test-appropriate practice materials. High-performance on this test is built on the bedrock of core learning and subject-matter proficiency. Deep content knowledge and plenty of practice are the keys to unlocking top scores.

This workbook contains over 6 exams' worth of questions – over 1,000 questions in total! We've taken a detailed look at the core content areas assessed on the Middle Level SSAT. We've created questions that will introduce students to these areas and improve their proficiency with practice. Our goal is to help students master these skills, increase their knowledge, and build up their confidence. To help with this, detailed answer explanations for every question are available online at **www.thetutorverse.com**.

Let's get started! You can use this workbook for independent study or with a professional teacher or tutor. Either way, we believe these learnings will benefit you on the Middle Level SSAT and beyond!

Good luck!

The Team at The Tutorverse

How to Use This Book

Overview

The purpose of this workbook is to provide students, parents, and educators with practice materials relevant to the Middle Level SSAT. This workbook assumes its users have a working knowledge of the exam, including its structure and content. Though it contains tips, strategies, and suggestions, the primary goal of this workbook is to provide students with extensive practice by introducing new words, skills, and concepts. A brief overview of the exam is shown below.

Scoring	Section	Number of Questions	Time Limit
Unscored Section (sent to schools)	Writing Sample	1	25 minutes
Scored Section	5-Minute Break		
	Section 1: Quantitative	25	30 minutes
	Section 2: Reading	40	40 minutes
	10-Minute Break		
	Section 3: Verbal	60	30 minutes
	Section 4: Quantitative	25	30 minutes
	Total Scored Exam (Sections 1-4)	**150**	**2 hours, 10 minutes**
Unscored Section	Section 5: Experimental	16	15 minutes

Organization

This workbook is organized into six main sections. Each section is designed to accomplish different objectives. These sections and objectives are as follows:

- Diagnostic Practice Test (Form A)
  The first full-length practice test is designed to help students identify the topics that require the most practice. It mirrors the length and content of the actual Middle Level SSAT in order to ensure that students become accustomed to the duration of the real test. This diagnostic practice test should be used to gauge the amount of additional practice needed on each topic, **not** as an estimate of how a student will score on the actual Middle Level SSAT. **NOTE:** the diagnostic practice test includes 16 questions in a mock-experimental section and, while they are useful practice, they are included only to emulate the full duration of the actual test.

- Quantitative
  The main concepts covered in this section are Number Concepts and Operations; Pre-Algebra; Algebra; Geometry; Measurement; Data Analysis; Statistics & Probability. All of these concepts are further divided into sub-categories, which can be found in the table of contents.

The Tutorverse
www.thetutorverse.com

- Reading

  This section tests a student's ability to read passages and answer questions about them. Passages include nonfiction persuasive and informative pieces, as well as excerpts from fictional works, including poems, novels, and short stories. Questions center around understanding main idea and themes, making inferences, and understanding how details contribute to the meaning of the passage.

- Verbal

  The material in this section covers word similarities and relationships through synonym and analogy questions. Students will encounter many new words in this section.

- Writing Sample

  This section provides information about the writing prompts (creative writing and essay-type), and includes several practice prompt pairs.

- Final Practice Test (Form B)

  This workbook ends with an additional full-length practice test. This test is similar to the diagnostic practice test in length and content and should be taken once students have completed the diagnostic practice test and have spent sufficient time answering the appropriate questions in the practice sections. **NOTE:** the practice test includes 16 questions in a mock-experimental section and, while they are useful practice, they are included only to emulate the full duration of the actual test.

Note: The Experimental section is designed by the SSAT Test Development Team to test new questions, in order to make sure they are appropriate for use on future SSAT exams. Since this section is **not** scored, this workbook will not include content related to the Experimental section (except in the Practice Test sections). Students need **not** worry about attempting these questions on the actual exam.

At the beginning of each of the above-listed sections are detailed instructions. Students should carefully review these instructions, as they contain important information about the actual exam and how best to practice.

Strategy

Every student has different strengths and abilities. We don't think there is any one strategy that will help every student ace the exam. Instead, we believe there are core principles to keep in mind when preparing for the Middle Level SSAT. These principles are interrelated and cyclical in nature.

- Evaluate

  A critical step in developing a solid study plan is to have a clear idea of how to spend your time. What subjects are more difficult for you? Which types of questions do you frequently answer incorrectly? Why? These and many other questions should be answered before developing any study plan. The diagnostic practice test is just one way to help you evaluate your abilities.

www.thetutorverse.com

- Plan
    Once you've taken stock of your strengths and abilities, focus on actions.
    - How much time do you have before the test?
    - How many areas do you need to work on during that time?
    - Which areas do you need to work on?
    - How many questions (and of which type) do you need to do each day, or each week?

    The answers to these and other questions will help you determine your study and practice plan.

- Practice
    Once you settle on a plan, try to stick with it as much as you can. To study successfully requires discipline, commitment, and focus. Try turning off your phone, TV, tablet, or other distractions. Not only will you learn more effectively when you're focused, but you may find that you finish your work more quickly, as well.

- Reevaluate
    Because learning and studying is an ongoing process, it is important to take stock of your improvements along the way. This will help you see how you are progressing and allow you to make adjustments to your plan. The practice test at the end of this workbook is designed to help you gauge your progress.

Need Help?

Feeling overwhelmed? You're not alone! Preparing for a standardized test is often a daunting task. While students should strive to meet the challenge of the test, it's also important for students to recognize when they need extra help.

Know that since the Middle Level SSAT is given to students in various grades, **students may find some material in this workbook difficult or entirely new**. It's OK! That is to be expected, as certain material may not have been taught to all students yet. Students will only be scored against other students in their grade (5th graders vs. other 5th graders, for example). Even so, mastering advanced materials often provides a competitive advantage in achieving higher scores. Give it a try!

Students are not alone. We provide detailed answer explanations online at **www.thetutorverse.com** (students should ask a parent or guardian's permission before going online). We also encourage students to reach out to trusted educators to help them prepare for the Middle Level SSAT. Experienced tutors, teachers, mentors, and consultants can help students with many aspects of their preparation – from evaluating and reevaluating their needs, to creating an effective plan to help them make the most of their practice.

Looking for a tutor?

Look no further – we're The Tutorverse for a reason! Have a parent or guardian send us an email at **hello@thetutorverse.com** and get started with a **free consultation**!

Diagnostic Practice Test (Form A)

Overview

The first step to an effective study plan is to determine a student's strengths and areas for improvement. This first practice test assesses a student's existing knowledge and grasp of concepts that may be seen on the actual exam.

Keep in mind that this practice test will be scored differently from the actual exam. On the actual Middle Level SSAT, **certain questions will not count towards a student's actual score (i.e. the experimental section)**. Also, the student's score will be determined by comparing his or her performance with those of other students in the same grade. On this practice test, however, every question is scored in order to accurately gauge the student's current ability level. Therefore, **this practice test should NOT be used as a gauge of how a student will score on the actual test**. This test should only be used to help students develop a study plan, and may be treated as a diagnostic test.

Format

The format of this diagnostic practice test is similar to that of the actual exam and includes 16 questions in a mock-experimental section. **For practice purposes only, students should treat the mock experimental section of the diagnostic practice test as any other.**

The format of the diagnostic practice test is below.

Scoring	Section	Number of Questions	Time Limit
Unscored Section (sent to schools)	Writing Sample	1	25 minutes
Scored Section	5-Minute Break		
	Section 1: Quantitative	25	30 minutes
	Section 2: Reading	40	40 minutes
	10-Minute Break		
	Section 3: Verbal	60	30 minutes
	Section 4: Quantitative	25	30 minutes
	Total Scored Exam (Sections 1-4)	150	2 hours, 10 minutes
Unscored Section	Section 5: Experimental	16	15 minutes

Answering

Use the answer sheet provided on the next page to record answers. Students may wish to tear it out of the workbook.

Diagnostic Practice Test (Form A)

Section 1: Quantitative

1. Ⓐ Ⓑ Ⓒ Ⓓ Ⓔ
2. Ⓐ Ⓑ Ⓒ Ⓓ Ⓔ
3. Ⓐ Ⓑ Ⓒ Ⓓ Ⓔ
4. Ⓐ Ⓑ Ⓒ Ⓓ Ⓔ
5. Ⓐ Ⓑ Ⓒ Ⓓ Ⓔ
6. Ⓐ Ⓑ Ⓒ Ⓓ Ⓔ
7. Ⓐ Ⓑ Ⓒ Ⓓ Ⓔ
8. Ⓐ Ⓑ Ⓒ Ⓓ Ⓔ
9. Ⓐ Ⓑ Ⓒ Ⓓ Ⓔ
10. Ⓐ Ⓑ Ⓒ Ⓓ Ⓔ
11. Ⓐ Ⓑ Ⓒ Ⓓ Ⓔ
12. Ⓐ Ⓑ Ⓒ Ⓓ Ⓔ
13. Ⓐ Ⓑ Ⓒ Ⓓ Ⓔ
14. Ⓐ Ⓑ Ⓒ Ⓓ Ⓔ
15. Ⓐ Ⓑ Ⓒ Ⓓ Ⓔ
16. Ⓐ Ⓑ Ⓒ Ⓓ Ⓔ
17. Ⓐ Ⓑ Ⓒ Ⓓ Ⓔ
18. Ⓐ Ⓑ Ⓒ Ⓓ Ⓔ
19. Ⓐ Ⓑ Ⓒ Ⓓ Ⓔ
20. Ⓐ Ⓑ Ⓒ Ⓓ Ⓔ
21. Ⓐ Ⓑ Ⓒ Ⓓ Ⓔ
22. Ⓐ Ⓑ Ⓒ Ⓓ Ⓔ
23. Ⓐ Ⓑ Ⓒ Ⓓ Ⓔ
24. Ⓐ Ⓑ Ⓒ Ⓓ Ⓔ
25. Ⓐ Ⓑ Ⓒ Ⓓ Ⓔ

Section 2: Reading

1. Ⓐ Ⓑ Ⓒ Ⓓ Ⓔ
2. Ⓐ Ⓑ Ⓒ Ⓓ Ⓔ
3. Ⓐ Ⓑ Ⓒ Ⓓ Ⓔ
4. Ⓐ Ⓑ Ⓒ Ⓓ Ⓔ
5. Ⓐ Ⓑ Ⓒ Ⓓ Ⓔ
6. Ⓐ Ⓑ Ⓒ Ⓓ Ⓔ
7. Ⓐ Ⓑ Ⓒ Ⓓ Ⓔ
8. Ⓐ Ⓑ Ⓒ Ⓓ Ⓔ
9. Ⓐ Ⓑ Ⓒ Ⓓ Ⓔ
10. Ⓐ Ⓑ Ⓒ Ⓓ Ⓔ
11. Ⓐ Ⓑ Ⓒ Ⓓ Ⓔ
12. Ⓐ Ⓑ Ⓒ Ⓓ Ⓔ
13. Ⓐ Ⓑ Ⓒ Ⓓ Ⓔ
14. Ⓐ Ⓑ Ⓒ Ⓓ Ⓔ
15. Ⓐ Ⓑ Ⓒ Ⓓ Ⓔ
16. Ⓐ Ⓑ Ⓒ Ⓓ Ⓔ
17. Ⓐ Ⓑ Ⓒ Ⓓ Ⓔ
18. Ⓐ Ⓑ Ⓒ Ⓓ Ⓔ
19. Ⓐ Ⓑ Ⓒ Ⓓ Ⓔ
20. Ⓐ Ⓑ Ⓒ Ⓓ Ⓔ
21. Ⓐ Ⓑ Ⓒ Ⓓ Ⓔ
22. Ⓐ Ⓑ Ⓒ Ⓓ Ⓔ
23. Ⓐ Ⓑ Ⓒ Ⓓ Ⓔ
24. Ⓐ Ⓑ Ⓒ Ⓓ Ⓔ
25. Ⓐ Ⓑ Ⓒ Ⓓ Ⓔ
26. Ⓐ Ⓑ Ⓒ Ⓓ Ⓔ
27. Ⓐ Ⓑ Ⓒ Ⓓ Ⓔ
28. Ⓐ Ⓑ Ⓒ Ⓓ Ⓔ
29. Ⓐ Ⓑ Ⓒ Ⓓ Ⓔ
30. Ⓐ Ⓑ Ⓒ Ⓓ Ⓔ
31. Ⓐ Ⓑ Ⓒ Ⓓ Ⓔ
32. Ⓐ Ⓑ Ⓒ Ⓓ Ⓔ
33. Ⓐ Ⓑ Ⓒ Ⓓ Ⓔ
34. Ⓐ Ⓑ Ⓒ Ⓓ Ⓔ
35. Ⓐ Ⓑ Ⓒ Ⓓ Ⓔ
36. Ⓐ Ⓑ Ⓒ Ⓓ Ⓔ
37. Ⓐ Ⓑ Ⓒ Ⓓ Ⓔ
38. Ⓐ Ⓑ Ⓒ Ⓓ Ⓔ
39. Ⓐ Ⓑ Ⓒ Ⓓ Ⓔ
40. Ⓐ Ⓑ Ⓒ Ⓓ Ⓔ

Section 3: Verbal

1. Ⓐ Ⓑ Ⓒ Ⓓ Ⓔ
2. Ⓐ Ⓑ Ⓒ Ⓓ Ⓔ
3. Ⓐ Ⓑ Ⓒ Ⓓ Ⓔ
4. Ⓐ Ⓑ Ⓒ Ⓓ Ⓔ
5. Ⓐ Ⓑ Ⓒ Ⓓ Ⓔ
6. Ⓐ Ⓑ Ⓒ Ⓓ Ⓔ
7. Ⓐ Ⓑ Ⓒ Ⓓ Ⓔ
8. Ⓐ Ⓑ Ⓒ Ⓓ Ⓔ
9. Ⓐ Ⓑ Ⓒ Ⓓ Ⓔ
10. Ⓐ Ⓑ Ⓒ Ⓓ Ⓔ
11. Ⓐ Ⓑ Ⓒ Ⓓ Ⓔ
12. Ⓐ Ⓑ Ⓒ Ⓓ Ⓔ
13. Ⓐ Ⓑ Ⓒ Ⓓ Ⓔ
14. Ⓐ Ⓑ Ⓒ Ⓓ Ⓔ
15. Ⓐ Ⓑ Ⓒ Ⓓ Ⓔ
16. Ⓐ Ⓑ Ⓒ Ⓓ Ⓔ
17. Ⓐ Ⓑ Ⓒ Ⓓ Ⓔ
18. Ⓐ Ⓑ Ⓒ Ⓓ Ⓔ
19. Ⓐ Ⓑ Ⓒ Ⓓ Ⓔ
20. Ⓐ Ⓑ Ⓒ Ⓓ Ⓔ
21. Ⓐ Ⓑ Ⓒ Ⓓ Ⓔ
22. Ⓐ Ⓑ Ⓒ Ⓓ Ⓔ
23. Ⓐ Ⓑ Ⓒ Ⓓ Ⓔ
24. Ⓐ Ⓑ Ⓒ Ⓓ Ⓔ
25. Ⓐ Ⓑ Ⓒ Ⓓ Ⓔ
26. Ⓐ Ⓑ Ⓒ Ⓓ Ⓔ
27. Ⓐ Ⓑ Ⓒ Ⓓ Ⓔ
28. Ⓐ Ⓑ Ⓒ Ⓓ Ⓔ
29. Ⓐ Ⓑ Ⓒ Ⓓ Ⓔ
30. Ⓐ Ⓑ Ⓒ Ⓓ Ⓔ
31. Ⓐ Ⓑ Ⓒ Ⓓ Ⓔ
32. Ⓐ Ⓑ Ⓒ Ⓓ Ⓔ
33. Ⓐ Ⓑ Ⓒ Ⓓ Ⓔ
34. Ⓐ Ⓑ Ⓒ Ⓓ Ⓔ
35. Ⓐ Ⓑ Ⓒ Ⓓ Ⓔ
36. Ⓐ Ⓑ Ⓒ Ⓓ Ⓔ
37. Ⓐ Ⓑ Ⓒ Ⓓ Ⓔ
38. Ⓐ Ⓑ Ⓒ Ⓓ Ⓔ
39. Ⓐ Ⓑ Ⓒ Ⓓ Ⓔ
40. Ⓐ Ⓑ Ⓒ Ⓓ Ⓔ
41. Ⓐ Ⓑ Ⓒ Ⓓ Ⓔ
42. Ⓐ Ⓑ Ⓒ Ⓓ Ⓔ
43. Ⓐ Ⓑ Ⓒ Ⓓ Ⓔ
44. Ⓐ Ⓑ Ⓒ Ⓓ Ⓔ
45. Ⓐ Ⓑ Ⓒ Ⓓ Ⓔ
46. Ⓐ Ⓑ Ⓒ Ⓓ Ⓔ
47. Ⓐ Ⓑ Ⓒ Ⓓ Ⓔ
48. Ⓐ Ⓑ Ⓒ Ⓓ Ⓔ
49. Ⓐ Ⓑ Ⓒ Ⓓ Ⓔ
50. Ⓐ Ⓑ Ⓒ Ⓓ Ⓔ
51. Ⓐ Ⓑ Ⓒ Ⓓ Ⓔ
52. Ⓐ Ⓑ Ⓒ Ⓓ Ⓔ
53. Ⓐ Ⓑ Ⓒ Ⓓ Ⓔ
54. Ⓐ Ⓑ Ⓒ Ⓓ Ⓔ
55. Ⓐ Ⓑ Ⓒ Ⓓ Ⓔ
56. Ⓐ Ⓑ Ⓒ Ⓓ Ⓔ
57. Ⓐ Ⓑ Ⓒ Ⓓ Ⓔ
58. Ⓐ Ⓑ Ⓒ Ⓓ Ⓔ
59. Ⓐ Ⓑ Ⓒ Ⓓ Ⓔ
60. Ⓐ Ⓑ Ⓒ Ⓓ Ⓔ

Section 4: Quantitative

1. Ⓐ Ⓑ Ⓒ Ⓓ Ⓔ
2. Ⓐ Ⓑ Ⓒ Ⓓ Ⓔ
3. Ⓐ Ⓑ Ⓒ Ⓓ Ⓔ
4. Ⓐ Ⓑ Ⓒ Ⓓ Ⓔ
5. Ⓐ Ⓑ Ⓒ Ⓓ Ⓔ
6. Ⓐ Ⓑ Ⓒ Ⓓ Ⓔ
7. Ⓐ Ⓑ Ⓒ Ⓓ Ⓔ
8. Ⓐ Ⓑ Ⓒ Ⓓ Ⓔ
9. Ⓐ Ⓑ Ⓒ Ⓓ Ⓔ
10. Ⓐ Ⓑ Ⓒ Ⓓ Ⓔ
11. Ⓐ Ⓑ Ⓒ Ⓓ Ⓔ
12. Ⓐ Ⓑ Ⓒ Ⓓ Ⓔ
13. Ⓐ Ⓑ Ⓒ Ⓓ Ⓔ
14. Ⓐ Ⓑ Ⓒ Ⓓ Ⓔ
15. Ⓐ Ⓑ Ⓒ Ⓓ Ⓔ
16. Ⓐ Ⓑ Ⓒ Ⓓ Ⓔ
17. Ⓐ Ⓑ Ⓒ Ⓓ Ⓔ
18. Ⓐ Ⓑ Ⓒ Ⓓ Ⓔ
19. Ⓐ Ⓑ Ⓒ Ⓓ Ⓔ
20. Ⓐ Ⓑ Ⓒ Ⓓ Ⓔ
21. Ⓐ Ⓑ Ⓒ Ⓓ Ⓔ
22. Ⓐ Ⓑ Ⓒ Ⓓ Ⓔ
23. Ⓐ Ⓑ Ⓒ Ⓓ Ⓔ
24. Ⓐ Ⓑ Ⓒ Ⓓ Ⓔ
25. Ⓐ Ⓑ Ⓒ Ⓓ Ⓔ

Section 5: Experimental

1. Ⓐ Ⓑ Ⓒ Ⓓ Ⓔ
2. Ⓐ Ⓑ Ⓒ Ⓓ Ⓔ
3. Ⓐ Ⓑ Ⓒ Ⓓ Ⓔ
4. Ⓐ Ⓑ Ⓒ Ⓓ Ⓔ
5. Ⓐ Ⓑ Ⓒ Ⓓ Ⓔ
6. Ⓐ Ⓑ Ⓒ Ⓓ Ⓔ
7. Ⓐ Ⓑ Ⓒ Ⓓ Ⓔ
8. Ⓐ Ⓑ Ⓒ Ⓓ Ⓔ
9. Ⓐ Ⓑ Ⓒ Ⓓ Ⓔ
10. Ⓐ Ⓑ Ⓒ Ⓓ Ⓔ
11. Ⓐ Ⓑ Ⓒ Ⓓ Ⓔ
12. Ⓐ Ⓑ Ⓒ Ⓓ Ⓔ
13. Ⓐ Ⓑ Ⓒ Ⓓ Ⓔ
14. Ⓐ Ⓑ Ⓒ Ⓓ Ⓔ
15. Ⓐ Ⓑ Ⓒ Ⓓ Ⓔ
16. Ⓐ Ⓑ Ⓒ Ⓓ Ⓔ

The Tutorverse
www.thetutorverse.com

Writing Sample

Schools would like to get to know you through an essay or story that you write. Choose one of the topics below that you find most interesting. Fill in the circle next to the topic of your choice. Then, write a story or essay based on the topic you chose.

Ⓐ If you could travel to a place you've never been before, where would you go and why?

Ⓑ It was my last chance.

Use this page and the next page to complete your writing sample.

SECTION 1
25 Questions

There are five suggested answers after each problem in this section. Solve each problem in your head or in the space provided to the right of the problem. Then, look at the suggested answers and pick the best one.

Note: Any figures or shapes that accompany problems in Section 1 are drawn as accurately as possible EXCEPT when it is stated that the figure is NOT drawn to scale.

Sample Question:

$11 \times 12 =$ ● Ⓑ Ⓒ Ⓓ Ⓔ

(A) 132
(B) 144
(C) 1,112
(D) 1,332
(E) 1,444

DO WORK IN THIS SPACE

1. Simplify the expression $(3 \times 4 - 3)^2 + 4$.
   (A) 13
   (B) 22
   (C) 27
   (D) 30
   (E) 85

2. 3, 8, 16, 27, 41, ___
   What is the next term in the series?
   (A) 14
   (B) 17
   (C) 55
   (D) 58
   (E) 61

3. 12, 1, −11, −2, −22, 2, 21
   What is the median of the list above?
   (A) −2
   (B) 0
   (C) 1
   (D) 2
   (E) 12

4. The radius of a circle is 1.4 cm. Using 3.14 for π, what is the approximate circumference of the circle? (*Note:* $C = \pi d$)
   (A) 4.396 cm
   (B) 6.154 cm
   (C) 8.792 cm
   (D) 43.96 cm
   (E) 87.92 cm

GO ON TO THE NEXT PAGE.

5. Richard eats $\frac{1}{2}$ of his birthday cake today. Tomorrow, he eats $\frac{1}{6}$ of the remaining cake. Which fraction represents how much cake is left?
   (A) $\frac{1}{12}$
   (B) $\frac{5}{12}$
   (C) $\frac{1}{6}$
   (D) $\frac{2}{3}$
   (E) $\frac{1}{3}$

6. How many quadrilaterals can be found in the shape shown at right?
   (A) 3
   (B) 4
   (C) 5
   (D) 6
   (E) 7

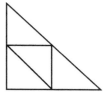

7. When Bash was born, he weighed 7.5 pounds. After one month, he weighed 9 pounds. By what percent did his weight increase from birth to one month?
   (A) 15%
   (B) $16\frac{2}{3}\%$
   (C) 20%
   (D) 25%
   (E) $33\frac{1}{3}\%$

8. One angle has a measure of 36°. A second angle has a measure of $b°$. Together, they form a right angle. What is the value of b?
   (A) 90
   (B) 72
   (C) 64
   (D) 54
   (E) 36

9. A group of 19 people share the bill for dinner. If the total bill was $734.06, what is the approximate cost for each person?
   (A) $10
   (B) $26
   (C) $37
   (D) $55
   (E) $60

10. 4 is the first number in a series. Each number after is double the preceding number. Which of the following is a number in the series?
    4, 8, 16, ...
    (A) 40
    (B) 80
    (C) 160
    (D) 256
    (E) 264

11. 53 × 47 − 2,000 =
    (A) 491
    (B) 583
    (C) 2,291
    (D) 4,471
    (E) 4,491

12. How many positive integers less than 50 are multiples of both 2 and 6?
    (A) 1
    (B) 2
    (C) 4
    (D) 6
    (E) 8

13. Tanisha has $\frac{3}{5}$ of an apple pie. She divides the pie among her 4 family members. What fraction of the pie will each person receive?
    (A) $\frac{1}{4}$
    (B) $\frac{1}{5}$
    (C) $\frac{1}{10}$
    (D) $\frac{3}{20}$
    (E) $\frac{9}{20}$

14. 34 × 46 − 36 =
    (A) 340
    (B) 680
    (C) 1,528
    (D) 1,558
    (E) 1,600

15. Round the number 0.42562 to the nearest thousandth.
    (A) 0.4
    (B) 0.42
    (C) 0.425
    (D) 0.4256
    (E) 0.426

GO ON TO THE NEXT PAGE.

16. Which of the following can NOT be drawn without lifting the pencil or retracting?
    (A)
    (B)
    (C)
    (D)
    (E)

17. 4 lions weigh as much as 3 tigers. 6 tigers weigh as much as 10 cheetahs. How many cheetahs weigh as much as 8 lions?
    (A) 6
    (B) 10
    (C) 15
    (D) 20
    (E) 24

18. If $5 - 2 \times j = 12$, what is the value of j?
    (A) -7
    (B) -4
    (C) $-\frac{7}{2}$
    (D) $\frac{7}{2}$
    (E) 4

19. If $KL = KM$, what is the perimeter of triangle KLM?
    (A) 11
    (B) 14
    (C) 15
    (D) 18
    (E) 28

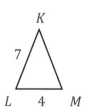

20. Which expression is equivalent to $2(4n + 3)$?
    (A) $4n + 6$
    (B) $8n + 3$
    (C) $8n + 6$
    (D) $14n$
    (E) $11n$

21. A garden has 2 rows of 12 carrots, 3 rows of 16 tomato plants, and 4 rows of 10 flowers. How many carrots, tomato plants, and flowers does the garden contain?
    (A) 47
    (B) 108
    (C) 112
    (D) 118
    (E) 342

22. In a jar, there are ten each of 1-cent, 5-cent, 10-cent, and 25-cent coins. If Teddy needs exactly 42 cents, what is the least number of coins he must take from the jar?
    (A) two
    (B) three
    (C) four
    (D) five
    (E) six

23. Solve for x:
    $-(x - 3) = 3$
    (A) −6
    (B) −3
    (C) 0
    (D) 3
    (E) 6

24. Gina is at the corner of Elm Street and Main Street as shown in the figure. If she walks two blocks west and three blocks south, she will be at:
    (A) point A
    (B) point B
    (C) point C
    (D) point D
    (E) point E

25. A point at (1, −7) is translated down and to the left. Which ordered pair represents a possible new location for the ordered point?
    (A) (1, −8)
    (B) (−1, −6)
    (C) (−3, −7)
    (D) (−4, −9)
    (E) (0, 1)

STOP
IF YOU FINISH BEFORE TIME IS UP,
CHECK YOUR WORK IN THIS SECTION ONLY.
YOU MAY NOT TURN TO ANY OTHER SECTION.

SECTION 2
40 Questions

Carefully read each passage and then answer the questions about it. For each question, select the choice that best answers the question based on the passage.

Ted quickly hurried away. He had no regrets about leaving behind the tedious task of farming. Now, he was going to be able to furnish a home for himself. How nice it would be to finally have a place which he could call his own! He could bring his books and his box of electrical things and settle down in his kingdom like a young lord. He did not
5   care if he only had a hammock to sleep in. No matter how tiny his room was, it made him happy to know he would be the king of his own realm. His imagination sped from one dream to another like lightning. If Mr. Wharton would let him run some wires from the barn to the shack, he would have electricity! He could light the room and heat it, too. He could even cook!
10   However, Mr. Wharton would probably consider such an idea out of the question. Ted would never be allowed to use the Wharton's electricity, of course. A candle would do for lighting, of course. Busy with these thoughts, Ted sped across the meadow and through the woods toward the river.

He soon came within sight of the shack on the water's edge. The shack had
15   molding walls, broken windows, and a caved-in roof. But for Ted Turner it possessed none of these bleak qualities. Instead of being a shack it seemed a paradise.

---

1. In lines 6-7, the sentence "His imagination sped from one dream to another like lightning" is an example of
   (A) simile
   (B) allegory
   (C) metaphor
   (D) personification
   (E) onomatopoeia

2. According to the passage, Mr. Wharton
   (A) is Ted's father
   (B) is very generous
   (C) never likes to share
   (D) has a barn near Ted's property
   (E) cares about Ted's well-being

3. Without changing the author's meaning, "bleak," line 16, could be replaced by
   (A) inevitable
   (B) gloomy
   (C) annoying
   (D) exciting
   (E) sharp

4. This passage is primarily about
   (A) the life of a former farmer
   (B) the various uses of electricity
   (C) a dangerous but valuable journey
   (D) the conflict between two characters
   (E) an exciting new development in a man's life

5. It can be inferred that Ted views the shack as a "paradise" (line 16) because he
   (A) is mentally unstable
   (B) grew up in a poor family
   (C) can finally cook with electricity
   (D) can keep many luxury items there
   (E) values freedom more than material things

GO ON TO THE NEXT PAGE.

Every year, the Department of Health (DOH) inspects New York City restaurants for compliance with food hygiene standards. The DOH awards the best-performing restaurants with an "A," while all others receive a "B" or "C." These health grades are then displayed prominently on the restaurant's façade for public view.

5   While there's no denying the importance of maintaining hygiene at an eating establishment, the current grading system is in urgent need of repair. For one thing, grades may have little to do with the hygiene of the food itself. Instead, the DOH may penalize restaurants for unrelated matters, like the storing of spoons in the same direction. Furthermore, grading is determined by a single inspector, who may or may not
10 be fair in his or her assessment.

Perhaps the biggest issue with the current grading system is a lack of transparency. Rules are often vague and confusing, making it difficult for restaurants to meet DOH standards. Since customers don't know how a grade was determined, they're unable to understand the significance of bad grades. Anything less than an "A" may
15 suggest rotting food or vermin, where a restaurant may have simply been penalized for having cracked floor tiles. Bad grades can drive away customers and be devastating to restaurant business.

To improve fairness, the DOH must revise its standards and better educate restaurants and customers alike.

6. As used in line 4, "façade" most nearly means
   (A) food
   (B) tables
   (C) exterior
   (D) kitchen
   (E) dining room

7. Which of the following best states the main idea of the third paragraph (lines 11-17)?
   (A) Customers should avoid restaurants with a "B" or "C" grades.
   (B) Restaurants with "B" or "C" grades have rats or cockroaches.
   (C) It is almost impossible for a restaurant to receive an "A."
   (D) Not knowing how health grades are determined is an unfair problem.
   (E) A health grade should be determined by more than one inspector.

8. According to the passage, restaurant grades are determined by
   (A) one person
   (B) online surveys
   (C) the mayor of the city
   (D) a team of health experts
   (E) restaurant patrons' reviews

9. The author would most likely agree with which of the following statements?
   (A) The DOH should be disbanded.
   (B) Having food hygiene standards is pointless.
   (C) Detailed food hygiene grading results should be made public.
   (D) The DOH should also include grades of "D" and "E" in its grading system.
   (E) Health inspection grades are private and should not be shown to customers.

10. Which of the following best describes the author's view of the DOH?
    (A) critical
    (B) admiring
    (C) impartial
    (D) encouraging
    (E) disappointed

GO ON TO THE NEXT PAGE.

In 1864, the United States government set aside land in the Yosemite Valley because of its natural beauty. Five years later, a young man by the name of John Muir arrived in the valley. Over the next few years, Muir worked, explored the valley, and wrote down his thoughts on nature. Muir's writings on the beauty and importance of
5 America's wild places would one day earn him the title "the father of the national parks."

After his time in Yosemite Valley, Muir wrote articles for magazines. He told readers that the United States was failing to manage its forests. Settlers in the American West had been cutting forests down too quickly. If it continued, lumber would become scarce. The United States government needed to write and enforce better laws. "God has
10 cared for these trees, saved them from drought, disease, and avalanches; but he cannot save them from fools – only Uncle Sam can do that," Muir wrote in 1897. President Theodore Roosevelt read Muir's work and wanted to meet him. In 1903, the two of them went camping together in Yosemite. While there, Muir urged Roosevelt to use his powers as president to protect more wild areas. Roosevelt eventually signed the Antiquities Act
15 of 1906, which allows the president to create protected national monuments.

Muir wanted people to become better caretakers of the natural world. "When we try to pick out anything by itself, we find it hitched to everything else in the Universe," Muir wrote. Many places around the world, like the polar ice caps and coral reefs, are in danger. We may not realize it, but the health of these ecosystems affects the health of
20 other ecosystems around the world. And like the forests of the American West in Muir's day, they will not recover unless we make stricter laws. We today should listen to Muir's message and start managing our effects on the environment better than we have so far.

11. The best title for this passage is
    (A) "John Muir's Travels"
    (B) "John Muir's Message"
    (C) "The Yosemite Valley"
    (D) "Muir and Roosevelt in Yosemite"
    (E) "The Father of the National Parks"

12. The phrase "the father of the national parks" (line 5) refers to
    (A) Uncle Sam
    (B) John Muir
    (C) Yosemite Valley
    (D) Theodore Roosevelt
    (E) settlers in the American West

13. The author most likely included the quote in lines 16–17 to
    (A) summarize Muir's life
    (B) provide a humorous example
    (C) prove that Muir is trustworthy
    (D) show that Muir's writing was interesting
    (E) show why caring for the natural world is important

14. The phrase "We may not realize it" in line 19 suggests that the author thinks
    (A) we don't know enough about the environment
    (B) polar ice caps and coral reefs are not very important
    (C) most people do not understand John Muir's message
    (D) most people are not concerned with places like the polar ice caps or coral reefs
    (E) the world needs another John Muir

15. The author's tone in the last paragraph is best described as
    (A) excited
    (B) concerned
    (C) admiring
    (D) curious
    (E) hopeless

This is the highest civilian honor this country can bestow, which is ironic, because nobody sets out to win it. But that's exactly what makes this award so special. What sets these men and women apart is the incredible impact they have had on so many people – not in short, blinding bursts, but steadily, over the course of a lifetime.

5   Bob Dylan started out singing other people's songs. But, as he says, "There came a point where I had to write what I wanted to say, because what I wanted to say, nobody else was writing." Born in Hibbing, Minnesota – a town, he says, where "you couldn't be a rebel – it was too cold" – Bob moved to New York at age 19. By the time he was 23, Bob's voice, with its weight, its unique, gravelly power, was redefining not just what music
10   sounded like, but the message it carried and how it made people feel. There is not a bigger giant in the history of American music. All these years later, he's still chasing that sound, still searching for a little bit of truth.

     Toni Morrison – she is used to a little distraction. As a single mother working at a publishing company by day, she would carve out a little time in the evening to write, often
15   with her two sons pulling on her hair and tugging at her earrings. Circumstances may not have been ideal, but the words that came out were magical. Toni Morrison's prose brings us that kind of moral and emotional intensity that few writers ever attempt. From "Song of Solomon" to "Beloved," Toni reaches us deeply, using a tone that is lyrical, precise, distinct, and inclusive. She believes that language "arcs toward the place where meaning
20   might lie." The rest of us are lucky to be following along for the ride.

     These are the recipients of the 2012 Medals of Freedom. I am proud to award them with this great honor.

16. According to the passage, what qualities were the recipients of the 2012 Medals of Freedom chosen for?
    (A) strong career ambitions
    (B) a lifelong devotion to their arts
    (C) inspirational musical talent
    (D) a belief in freedom as a basic human right
    (E) being good parents during tough times

17. The word "bestow" (line 1) most likely means
    (A) hide
    (B) imagine
    (C) stowaway
    (D) grant
    (E) believe in

18. The idea that language "arcs toward the place where meaning might lie" (lines 19-20) means
    (A) writing is a way to search for truth
    (B) literature often fails to be meaningful
    (C) Toni Morrison is a great poet
    (D) language is full of motion
    (E) words can be as beautiful as a rainbow

19. The passage indicates that Bob Dylan left his hometown because
    (A) he went to college in New York
    (B) he would get sick from the cold weather
    (C) he wanted to redefine what rock music could sound like
    (D) nobody in his hometown appreciated his vocal talents
    (E) he wanted to rebel and express himself

20. Based on the tone of the speech, how does the author feel about Bob Dylan and Toni Morrison?
    (A) amused and entertained
    (B) disappointed in the judging decision
    (C) envious of their talent
    (D) deeply admiring
    (E) confused by their work

GO ON TO THE NEXT PAGE.

Many historians have a dim view of Herbert Hoover's presidency. They argue that his leadership prolonged the Great Depression, a period between 1929 and the 1940 when one fourth of the nation's population lost their jobs and many became homeless. Historians claim that Hoover should have done more to relieve the economic disaster
5 facing the country, yet these arguments fail to highlight how several of Hoover's actions actually did help alleviate America's greatest economic crisis.

Many historians overlook Hoover's Federal Home Bank Act of 1932. This act helped to stabilize homeownership in the United States. It made it easier for the average American to borrow money to buy a home. It also created a commercial safety net for real
10 estate by setting aside $125 million dollars for home insurance. Hoover saw that homeownership was an important part of the American economy.

Hoover was not as unresponsive as many of his opponents claimed. He tried to create jobs with the Glass-Steagall Act and the Emergency Relief and Construction Act in 1932. These acts used government gold reserves to help struggling businesses, fund
15 railroad construction projects, and create new job opportunities for struggling Americans. Along with the Federal Home Bank Act, these acts helped ease the troubles of the Great Depression.

Though Hoover's loss in the next presidential election meant that his policies did not have time to reach their full potential impact, his decisions did help make a difference.
20 He was a man who inherited a difficult situation, not the lame duck historians often make him out to be. After all, his policies helped set the country on the path to recovery, and his intervention efforts set the stage for the New Deal order of the next several decades.

21. The tone of the passage is best described as
   (A) emotional
   (B) persuasive
   (C) critical
   (D) factual
   (E) hopeful

22. The use of the phrase "lame duck" (line 20) shows that historians felt
   (A) angry at Hoover's support staff
   (B) accepting of Hoover's limitations
   (C) disdainful of Hoover's efforts
   (D) approving of Hoover's policies
   (E) uninterested in Hoover's decisions

23. The central idea of the passage is that
   (A) Herbert Hoover was a root cause of the Great Depression
   (B) Herbert Hoover was not as inactive during the Great Depression as many people claim
   (C) Herbert Hoover inherited the Great Depression and could not change it
   (D) Herbert Hoover saw homeownership as the beating heart of the American Dream
   (E) Herbert Hoover deserved to be re-elected during the Great Depression

24. The word "alleviate" (line 6) most likely means
   (A) destroy
   (B) allow
   (C) relieve
   (D) strengthen
   (E) worsen

25. According to the passage, why did Herbert Hoover's policies "not have time to reach their full potential impact" (line 19)?
   (A) He changed his policies too soon after implementing them.
   (B) Historians and politicians prevented his policies from being realized.
   (C) He prolonged the Great Depression.
   (D) He was not re-elected for presidency.
   (E) He focused too many resources on the New Deal, and the gold reserves ran out.

>   Above the east horizon,
>   The great red flower of the dawn
>   Opens slowly, petal by petal;
>   The trees emerge from darkness
> 5 With ghostly silver leaves,
>   Dew powdered.
>   Now consciousness emerges
>   Reluctantly out of tides of sleep;
>   Finding with cold surprise
> 10 No strange new thing to match its dreams,
>   But merely the familiar shapes
>   Of bedpost, window-pane, and wall.
>
>   Within the city,
>   The streets which were the last to fall to sleep,
> 15 Hold yet stale fragments of the night.
>   Sleep oozes out of stagnant ash-barrels,
>   Sleep drowses over litter in the streets.
>   Sleep nods upon the milkcans by back doors.
>   And, in shut rooms,
> 20 Behind the lowered window-blinds,
>   Drawn white faces unwittingly flout the day.
>
>   But, at the edges of the city,
>   Sleep is already washed away;
>   Light filters through the moist green leaves,
> 25 It runs into the cups of flowers,
>   It leaps in sparks through drops of dew,
>   It whirls against the window-panes
>   With waking birds;
>   Blinds are rolled up and chimneys smoke,
> 30 Feet clatter past in silent paths,
>   And down white vanishing ways of steel,
>   A dozen railway trains converge
>   Upon night's stronghold.

26. The mood of the poem is best described as
    (A) sly
    (B) awake
    (C) fearful
    (D) straightforward
    (E) neutral

27. It can be inferred that the "edges of the city" (line 22) are
    (A) silent
    (B) stale
    (C) busy
    (D) reluctant
    (E) sleepy

28. The figure of speech represented in line 2 is
    (A) simile
    (B) metaphor
    (C) allusion
    (D) personification
    (E) alliteration

29. Which of the following best describes the main idea of the second stanza of the poem?
    (A) The morning shines brightly within the city.
    (B) Drawn white faces are excited for the new day.
    (C) Sleepy people wake up by raising their window-blinds.
    (D) Many city residents only go to sleep when the sun comes up.
    (E) The streets within the city are busy with partygoers.

30. The word "flout," as used in line 21, most likely means
    (A) welcome
    (B) save
    (C) enjoy
    (D) dismiss
    (E) capture

People should be more concerned about their daily water intake. Water is essential for the healthy functioning of the human body, and makes up 70% of a person's body weight. Because of this, it is important for people to drink water regularly throughout the day.

5   When the body does not have enough water, it can become dehydrated. This condition can be caused by illness, exercise, heat or even stress. Dehydration can be very damaging. It can be the source of a simple headache, or it can lead a person to collapse or faint. Even worse, it can cause the body to overheat or even result in a heart attack.

It's sometimes difficult to tell, but the body is constantly losing water. If a person
10 waits to hydrate until she is thirsty, then it is already too late. This is because thirst is a delayed sign of dehydration. The body's thirst mechanism fails to "notify" the body in time. When thirst sets in, dehydration may already be at work damaging the kidneys and other organs.

These unpleasant effects can be prevented by drinking water throughout the day.
15 How much water a person needs varies based on their height, weight, activity, and metabolism. Generally, however, a person should drink between 2.7 and 3.7 liters of water each day. Some experts suggest that 80 percent of a person's water should come from fluids, while the other 20 percent should come from foods like fruits and vegetables.

By having a balanced diet and being aware of fluid intake, a person can easily
20 avoid the pernicious effects of dehydration.

31. The passage can be best described as
    (A) an epic poem
    (B) a fictional story
    (C) a persuasive essay
    (D) a product advertisement
    (E) an entry from a personal journal

32. Which of the following conclusions can be drawn from the passage?
    (A) It is easy to stay hydrated.
    (B) The majority of your water intake should come from food.
    (C) Water accounts for a small percentage of your body weight.
    (D) Thirst is an immediate sign of dehydration.
    (E) Damage from dehydration is not always outwardly visible.

33. It can be inferred that the word "pernicious" (line 20) means
    (A) balanced
    (B) harmful
    (C) beneficial
    (D) delayed
    (E) favorable

34. Which of the following is the author most likely to discuss next?
    (A) different strategies to help stay hydrated
    (B) reasons why water is better for hydration than fruits and vegetables
    (C) ways the kidneys and other organs are affected by dehydration
    (D) the amount of water people should drink each day
    (E) the causes of dehydration

35. The headline that best fits the passage is
    (A) "Dehydration as Leading Cause of Kidney Failure and Heart Attacks"
    (B) "The Importance of Drinking Water"
    (C) "The Body's Delayed Thirst Mechanism"
    (D) "How to Stay Hydrated"
    (E) "The Role of Fruits and Vegetables in Fighting Dehydration"

GO ON TO THE NEXT PAGE.

It was the day when Mother had gone to Maidbridge. She had gone alone, but the children planned to go to the station to meet her. Loving the station as they did, it was only natural that the children should be there a good hour before there was any chance of Mother's train arriving. They would be early even if the train was punctual, which was
5 most unlikely. No doubt they would have been just as early if it had been a fine day, and all the delights of woods and fields and rocks and rivers had been open to them. But it happened to be a very wet day and very cold for July. There was a wild wind that drove flocks of dark purple clouds across the sky "like herds of dream-elephants," as Phyllis said. And the rain stung sharply, so that the walk to the station was finished with a run.
10 Then the rain fell faster and harder, and beat slantwise against the windows of the booking office and the door of the General Waiting Room.
"It's like being in a besieged castle," Phyllis said; "look at the arrows of the foe striking against the battlements!"
"It's much more like a great garden-squirt," said Peter.
15 They decided to wait on the up side, for the down platform looked very wet indeed. They saw that the rain was driving right into the little bleak shelter where down-passengers wait for their trains.
The hour would be full of incident and of interest, for there would be two up trains and one down to look at before the one that should bring Mother back.
20 "Perhaps it'll have stopped raining by then," said Bobbie. "Anyhow, I'm glad I brought Mother's waterproof and umbrella."

36. As used in line 4, the word "punctual" most nearly means
    (A) authentic
    (B) broken
    (C) noisy
    (D) prompt
    (E) comfortable

37. It can be inferred from the passage that the "battlements" in line 13 refer to the
    (A) Maidbridge houses
    (B) cloudy and dark sky
    (C) castle in the distance
    (D) booking office and waiting room
    (E) trains passing through the station

38. The main reason the children run to the station is that they
    (A) do not have umbrellas
    (B) are afraid of being late
    (C) find the rain uncomfortable
    (D) cannot contain their excitement
    (E) have been instructed to do so by their mother

39. According to the passage, if the weather was better, the children would have
    (A) stayed at home
    (B) not changed their plans
    (C) missed their mother's train
    (D) ran the entire way to the station
    (E) taken a scenic route to the station

40. The passage suggests that the children will enjoy waiting at the station because they
    (A) have nothing else to do
    (B) will be able to stay dry
    (C) are excited to see their mother
    (D) love watching the activity in the station
    (E) rarely have a chance to be on their own

STOP
**IF YOU FINISH BEFORE TIME IS UP,
CHECK YOUR WORK IN THIS SECTION ONLY.
YOU MAY NOT TURN TO ANY OTHER SECTION.**

SECTION 3
60 Questions

There are two different types of questions in this section: synonyms and analogies. Read the directions and sample question for each type.

Synonyms

Each of the questions that follow consist of one capitalized word. Each word is followed by five words or phrases. Select the one word or phrase whose meaning is closest to the word in capital letters.

Sample Question:

1. PUBLISH:
   (A) journalism
   (B) edit
   (C) book
   (D) suspend
   (E) print

2. ACCOMPANY:
   (A) business
   (B) join
   (C) concern
   (D) discover
   (E) friend

3. NORMAL:
   (A) compact
   (B) usually
   (C) direct
   (D) associate
   (E) regular

4. CULTURE:
   (A) rich
   (B) customs
   (C) intelligence
   (D) music
   (E) teach

5. ATTACH:
   (A) melt
   (B) repair
   (C) conjoin
   (D) bestow
   (E) attachment

6. BEHAVIOR:
   (A) naughty
   (B) accent
   (C) anniversary
   (D) adjustment
   (E) attitude

7. PREOCCUPATION:
   (A) obsessed
   (B) seek
   (C) conduct
   (D) fixation
   (E) job

8. APPRECIATE:
   (A) topical
   (B) gift
   (C) restrict
   (D) occur
   (E) thank

GO ON TO THE NEXT PAGE.

9. SCHEDULE:
   - (A) reservation
   - (B) timetable
   - (C) income
   - (D) holiday
   - (E) assumption

10. ADMISSION:
    - (A) theater
    - (B) ticket
    - (C) intent
    - (D) entrance
    - (E) previous

11. LAYER:
    - (A) object
    - (B) warmth
    - (C) coating
    - (D) bounty
    - (E) descriptive

12. BEWILDERING:
    - (A) confusing
    - (B) wilderness
    - (C) grant
    - (D) abnormal
    - (E) domestic

13. CHART:
    - (A) information
    - (B) pie
    - (C) conclude
    - (D) captain
    - (E) diagram

14. ASSIGN:
    - (A) designate
    - (B) adjust
    - (C) paperwork
    - (D) allow
    - (E) homework

15. DEFICIENCY:
    - (A) lack
    - (B) tense
    - (C) proper
    - (D) missing
    - (E) vitamin

16. LIKEWISE:
    - (A) minimum
    - (B) agreement
    - (C) almost
    - (D) discreet
    - (E) similarly

17. COMPULSION:
    - (A) small desire
    - (B) wishful thinking
    - (C) random pulsing
    - (D) rich context
    - (E) suggested habit

18. AUTHOR:
    - (A) clerk
    - (B) autograph
    - (C) writer
    - (D) book
    - (E) library

19. DOCTRINE:
    - (A) saying
    - (B) lenient
    - (C) healthy
    - (D) invented
    - (E) bathroom

20. FICTITIOUS:
    - (A) made up
    - (B) most generous
    - (C) not responsible
    - (D) entertaining
    - (E) extremely thankful

21. CODE:
    - (A) create
    - (B) occupy
    - (C) bank
    - (D) password
    - (E) enter

22. DELIBERATE:
    - (A) expert
    - (B) argue
    - (C) intentional
    - (D) jury
    - (E) only

GO ON TO THE NEXT PAGE.

23. TEXT:
    (A) handy
    (B) delete
    (C) words
    (D) friends
    (E) phone

24. SURVEY:
    (A) result
    (B) state
    (C) poll
    (D) answer
    (E) source

25. ABSENCE:
    (A) vacancy
    (B) student
    (C) disappear
    (D) beloved
    (E) school

26. COMPENSATION:
    (A) necessary
    (B) payment
    (C) happiness
    (D) judgment
    (E) promotion

27. CONSIDERABLE:
    (A) kind
    (B) large
    (C) polite
    (D) considerate
    (E) deafening

28. PREJUDICE:
    (A) fair
    (B) juicy
    (C) curious
    (D) prevention
    (E) preconception

29. BESEECH:
    (A) wonder
    (B) explain
    (C) beg
    (D) unravel
    (E) speak

30. UNPREDICTABLE:
    (A) unstable
    (B) achievable
    (C) abduct
    (D) guessing
    (E) knowable

GO ON TO THE NEXT PAGE.

Analogies

The questions that follow ask you to find relationships between words. For each question, select the answer choice that best completes the meaning of the sentence.

Sample Question:

> Jump is to leap as: ●⒝©⒟⒠
>
> (A) twirl is to spin
> (B) dance is to dancer
> (C) runner is to race
> (D) hot is to cold
> (E) happy is to sad

Choice (A) is the best answer because jump and leap are synonyms, just as twirl and spin are synonyms. This choice states a relationship that is most like the relationship between jump and leap.

31. Birth is to life as
    (A) clock is to wrist
    (B) painter is to script
    (C) neglect is to shabbiness
    (D) soda is to can
    (E) colleague is to coworker

32. Salute is to officer as
    (A) hastily is to reluctantly
    (B) castle is to attack
    (C) envision is to future
    (D) credibility is to believable
    (E) refer is to search

33. Teammate is to team as
    (A) foe is to against
    (B) ferment is to grow
    (C) seed is to watermelon
    (D) uniform is to costume
    (E) action is to instruction

34. Rain is to downpour as
    (A) strong is to almighty
    (B) brave is to bravery
    (C) fee is to payment
    (D) part is to segment
    (E) estimate is to guess

35. Fluent is to inarticulate as
    (A) fall is to tumble
    (B) surprise is to gift
    (C) encounter is to happen
    (D) massive is to minute
    (E) enchanting is to mysterious

36. Glue is to bind as
    (A) model is to inspiration
    (B) illusion is to sense
    (C) structure is to emphasize
    (D) margin is to ruler
    (E) axe is to hack

37. Photographer is to camera as
    (A) bet is to gambler
    (B) hermit is to alone
    (C) banker is to teller
    (D) musician is to instrument
    (E) lawyer is to policeman

38. Mirror is to reflection as
    (A) shade is to chilly
    (B) sanitize is to clean
    (C) polish is to gleam
    (D) point is to sharp
    (E) beard is to trim

GO ON TO THE NEXT PAGE.

39. Textbook is to study as
    (A) instinct is to natural
    (B) ladder is to climb
    (C) budget is to limitation
    (D) rate is to speed
    (E) reconcile is to replace

40. Breeze is to squall as
    (A) squeeze is to twist
    (B) courageous is to powerful
    (C) knowledgeable is to omniscient
    (D) sigh is to exhale
    (E) expectation is to await

41. Announcer is to amplifier as stylist is to
    (A) hairstyle
    (B) imagination
    (C) cut
    (D) taste
    (E) comb

42. Morose is to gloomy as tremendous is to
    (A) power
    (B) experience
    (C) slight
    (D) immense
    (E) wealth

43. Illustrate is to illustrated as
    (A) apathy is to apathetic
    (B) formed is to form
    (C) emotion is to emotional
    (D) anticipate is to anticipated
    (E) extra is to extraneous

44. Bird is to egg as
    (A) lantern is to light
    (B) moon is to formation
    (C) bulb is to watt
    (D) feather is to fur
    (E) sticker is to collection

45. Career is to engineering as
    (A) monitor is to spy
    (B) language is to English
    (C) founder is to proprietor
    (D) jungle is to wilderness
    (E) hat is to accessory

46. Insignificant is to unimportant as
    (A) legion is to few
    (B) assault is to hug
    (C) falter is to stabilize
    (D) phonic is to spoke
    (E) malign is to malevolent

47. Hotel is to guests as barn is to
    (A) home
    (B) outside
    (C) hut
    (D) livestock
    (E) safety

48. Elated is to miserable as
    (A) inquisitive is to irritable
    (B) cliff is to ascend
    (C) fluid is to horrid
    (D) decade is to ten
    (E) permanent is to temporary

49. Luckiest is to lucky as
    (A) hungriest is to hungry
    (B) greatly is to great
    (C) carefully is to care
    (D) assembly is to assemble
    (E) fortunate is to unfortunate

50. Quilt is to patch as
    (A) highway is to lane
    (B) role is to memorize
    (C) headlight is to gear
    (D) heirloom is to keepsake
    (E) weave is to sew

51. Plan is to meeting as
    (A) gather is to inclusive
    (B) deliver is to package
    (C) delusion is to fantasy
    (D) hospitable is to surgeon
    (E) reprimand is to superior

52. Inspiration is to idea as emergency is to
    (A) dire
    (B) hospital
    (C) medic
    (D) panic
    (E) patient

GO ON TO THE NEXT PAGE.

53. Megaphone is to protest as
    (A) congestion is to clogged
    (B) box is to contain
    (C) zipper is to bundle
    (D) bag is to back
    (E) knob is to spin

54. Typist is to keyboard as
    (A) marker is to calendar
    (B) villain is to evil
    (C) studio is to producer
    (D) singer is to microphone
    (E) geologist is to dug

55. Inexcusable is to unforgiveable as
    (A) evade is to taxes
    (B) diligent is to lazy
    (C) serial is to crime
    (D) retort is to witty
    (E) partial is to biased

56. Angry is to furious as sad is to
    (A) annoyed
    (B) upset
    (C) mad
    (D) overjoyed
    (E) inconsolable

57. Seasoning is to enhance as
    (A) chop is to mince
    (B) salt is to pepper
    (C) savory is to bland
    (D) spice is to flavor
    (E) boiled is to sautéed

58. Insect is to mosquito as
    (A) general is to detailed
    (B) container is to box
    (C) height is to width
    (D) bowl is to dish
    (E) math is to geometry

59. King is to country as
    (A) mayor is to city
    (B) governor is to world
    (C) Congress is to Parliament
    (D) parent is to grandparent
    (E) barista is to coffee

60. Woman is to women as man is to
    (A) manly
    (B) geese
    (C) feminine
    (D) boyish
    (E) men

STOP
IF YOU FINISH BEFORE TIME IS UP,
CHECK YOUR WORK IN THIS SECTION ONLY.
YOU MAY NOT TURN TO ANY OTHER SECTION.

SECTION 4
25 Questions

There are five suggested answers after each problem in this section. Solve each problem in your head or in the space provided to the right of the problem. Then look at the suggested answers and pick the best one.

Note: Any figures or shapes that accompany problems in Section 1 are drawn as accurately as possible EXCEPT when it is stated that the figure is NOT drawn to scale.

Sample Question:

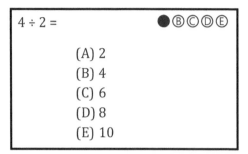

DO WORK IN THIS SPACE

1. Which phrase represents the expression $5(10 - n)$?
   (A) the product of five, ten, and n
   (B) five times ten divided by n
   (C) five times ten minus n
   (D) five times the product of ten and n
   (E) five times the difference of ten and n

2. Mrs. Smith is 4 times Maya's age, and Bob is 2 years older than Maya. If Mrs. Smith is 36 years old, how old is Bob?
   (A) 9
   (B) 11
   (C) 18
   (D) 30
   (E) 34

3. In 1980, the population of Little Falls was 12,000. In 2010, the population was 8,000. By what percent did the population decrease from 1980 to 2010?
   (A) $33\frac{1}{3}\%$
   (B) 40%
   (C) 50%
   (D) $66\frac{2}{3}\%$
   (E) 80%

GO ON TO THE NEXT PAGE.

4. Three ivory tiles can be traded for 4 pearls. Six pearls can be traded for 7 garnets. How many garnets can be traded for 36 ivory tiles?
   (A) 12
   (B) 14
   (C) 28
   (D) 42
   (E) 56

5. What is the correct order, from least to greatest, of the following series of numbers?
   0.02, 0.015, 0.34, 0.0083, 0.4
   (A) 0.015, 0.02, 0.34, 0.4, 0.0083
   (B) 0.02, 0.015, 0.34, 0.4, 0.0083
   (C) 0.0083, 0.02, 0.015, 0.34, 0.4
   (D) 0.0083, 0.015, 0.02, 0.34, 0.4
   (E) 0.4, 0.34, 0.02, 0.015, 0.0083

6. Bonnie's Ice Cream Shoppe has the flavors and cone options shown in the table to the right. How many different orders can be made if each order can have one flavor and one cone option?
   (A) three
   (B) five
   (C) eight
   (D) eleven
   (E) fifteen

Flavors	Cone
Chocolate	Regular
Vanilla	Sugar
Cherry	Waffle
Cookie Dough	
Rocky Road	

7. Which expression has the greatest value, if $m = 1$ and $n = 3$?
   (A) $m + n$
   (B) mn
   (C) mn^2
   (D) $m + n^2$
   (E) $\frac{m}{n}$

8. In a recent election between three candidates, Alicia received 35,000 votes, and Ming received three times as many votes as Alicia. If 250,000 votes were cast, and every voter chose one of the candidates, how many votes did Marco, the third candidate, receive?
   (A) 110,000
   (B) 140,000
   (C) 150,000
   (D) 210,000
   (E) 215,005

9. When *x* and *y* are positive consecutive integers, which statement must be true?
   (A) The value of *x* + *y* must be odd
   (B) The value of *x* − *y* must be positive
   (C) The value of *xy* must be odd
   (D) The value of $\frac{x}{y}$ must be greater than 1
   (E) The value of $\frac{y}{x}$ must be less than 1

10. Dan is paid 32 dollars for working 3 hours. If Dan works 9 hours on Monday and a total of 12 hours combined on Tuesday and Wednesday, how much did he earn for the three days?
    (A) $96
    (B) $224
    (C) $256
    (D) $288
    (E) $352

11. A food stand sells hot dogs, soda, and pretzels. A hot dog costs $2.50, a cup of soda costs $1.50, and a pretzel costs $0.75. The Reilly family bought 2 hot dogs and 3 pretzels. If they spent $13.25 in total, how many cups of soda did they buy?
    (A) 0
    (B) 2
    (C) 3
    (D) 4
    (E) 6

12. 5,432 + 6,789 − 3,443 =
    (A) 8,888
    (B) 8,878
    (C) 8,788
    (D) 8,778
    (E) 8,768

13. Jason is between $50\frac{1}{6}$ and $50\frac{1}{2}$ inches tall. Which could be Jason's height, in inches?
    (A) 50.1
    (B) 50.15
    (C) 50.25
    (D) 50.65
    (E) 50.8

Students at Greenwood Middle School were asked which mode of transportation they used to get to school each day. To the right is a circle graph displaying the results.

Modes of Transportation

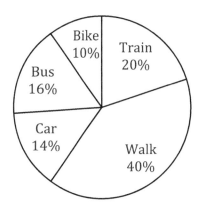

14. If 450 students were surveyed, how many students walk to school?
    (A) 40
    (B) 45
    (C) 180
    (D) 270
    (E) 400

15. Mrs. Daisy's kindergarten class has 1 gallon of regular milk and 1 gallon of chocolate milk. Two students each drink $\frac{1}{3}$ of the regular milk and 4 students each drink $\frac{1}{5}$ of the chocolate milk. How much combined milk, in gallons, did Mrs. Daisy's class drink?
    (A) $\frac{8}{15}$
    (B) $1\frac{3}{5}$
    (C) $1\frac{7}{15}$
    (D) $1\frac{13}{15}$
    (E) $1\frac{19}{20}$

16. The 10th grade class at Grace Hopper High School has 150 students. Each student takes one language course: Spanish, French, or German. If $\frac{1}{5}$ take French, and $\frac{1}{3}$ take Spanish, how many students combined take either French or Spanish?
    (A) 30
    (B) 50
    (C) 70
    (D) 80
    (E) 100

GO ON TO THE NEXT PAGE.

17. The graph to the right shows the number of plants in a garden. How many more tomato plants than squash plants are in the garden?
    (A) 2
    (B) 7
    (C) 8
    (D) 9
    (E) 10

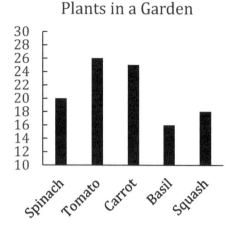

18. A student surveys her classmates for a science project, asking them how much time they spend on social media and how much sleep they get every night. Using the graph to the right, how many more hours of sleep do students who use social media for three hours get than those who use social media for five hours?
    (A) 0
    (B) 1
    (C) 2
    (D) 3
    (E) 4

19. There are 40 cans of soda in the stockroom of a deli. Of these cans of soda, $\frac{1}{8}$ are orange soda. Which expression can be used to find how many cans are orange soda?
    (A) $40 \div \frac{1}{8}$
    (B) $40 + \frac{1}{8}$
    (C) $40 - \frac{1}{8}$
    (D) $40 \times \frac{1}{8}$
    (E) $\frac{1}{8} \div 40$

20. A square piece of paper has an area of 100 cm². What is its perimeter?
    (A) 10 cm
    (B) 20 cm
    (C) 40 cm
    (D) 100 cm
    (E) 400 cm

GO ON TO THE NEXT PAGE.

21. What is the area, in square units, of the rectangle shown to the right?
    (A) 30
    (B) 32
    (C) 42
    (D) 60
    (E) 72

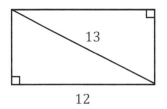

22. 936 ÷ 12 =
    (A) 88
    (B) 83
    (C) 78
    (D) 73
    (E) 68

23. The circumference of a circle is 21.98 meters. What is the approximate radius of the circle? (*Note: π = 3.14 and C = πd*)
    (A) 3.5m
    (B) 7m
    (C) 12.25m
    (D) 14m
    (E) 49m

24. Which expression has the least value?
    (A) 2,395 ÷ 100
    (B) 2,395 ÷ 10
    (C) 23.95 × 0.1
    (D) 239.5 × 10
    (E) 2,395 × 0.01

Use the graph to the right to solve the question.

25. If 4,000 people live in Fort Riley, how many are older than 60 and younger than 81?
    (A) 20
    (B) 200
    (C) 400
    (D) 800
    (E) 1,000

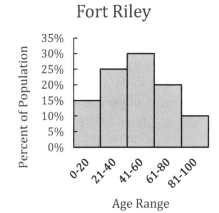

STOP
IF YOU FINISH BEFORE TIME IS UP,
CHECK YOUR WORK IN THIS SECTION ONLY.
YOU MAY NOT TURN TO ANY OTHER SECTION.

SECTION 5
16 Questions

Today, Bob Dylan is widely recognized as a leader of the American folk music genre. Many consider songs like "Blowin' in the Wind" and "The Times They Are a-Changin'" to have defined the genre itself. Yet, Dylan was not always hailed as the King of Folk.

5   In 1965, the musician did something very controversial. He decided to "go electric" at the Newport Folk Festival. Instead of playing with his traditional acoustic guitar, Dylan decided to play an electric guitar. There are different accounts of what happened at of the festival that day. However, many attendees agree that Dylan's switch in instruments triggered boos from the audience.

10   Today, it is not uncommon for artists to use different instruments. However, during the 1960s, most people associated the acoustic guitar with folk music, and the electric guitar with rock n' roll. At the time, rock n' roll was still getting it's start, and was a very progressive genre. When concert-goers saw Dylan playing a rock n' roll instrument, they were frustrated that Dylan appeared to be abandoning the customary

15   elements of folk music. Others worried that Dylan's decision would cause the growing popularity of folk music to come to a screeching halt.

1. As used in line 3, the word "hailed" most nearly means
   (A) rained
   (B) waved
   (C) stormed
   (D) assumed
   (E) acknowledged

2. In lines 5-6, the author uses the phrase "go electric" to refer to
   (A) rock n' roll's growing popularity
   (B) Dylan's popularity during the 1960s
   (C) the success of "Blowin' in the Wind"
   (D) Dylan's choice to change instruments
   (E) the booing audience at the Newport Folk Festival

3. According to the passage,
   (A) Dylan never intended to play an electric guitar
   (B) Dylan did not care about being booed by an audience
   (C) Dylan wished to be seen as progressive, not traditional
   (D) "Blowin' in the Wind" is the most famous folk song of all time
   (E) not everyone agrees about what happened at the 1965 Newport Folk Festival

4. The author implies which of the following about folk music?
   (A) It did not gain much attention in the 1960s.
   (B) It helped popularize the electric guitar.
   (C) It was more traditional than rock n' roll music.
   (D) It attracted many critics who did not like the style.
   (E) It remains one of the most popular music genres.

5. The passage implies that, as a musician, Bob Dylan was
   (A) very influential
   (B) underappreciated
   (C) unhappy with folk music
   (D) regarded as a rock n' roll icon
   (E) more concerned with money than music

6. SYMPATHY:
   (A) emotion
   (B) sorry
   (C) progress
   (D) apology
   (E) kindness

GO ON TO THE NEXT PAGE.

7. COMPLICATION:
   (A) obstacle
   (B) difficult
   (C) obedience
   (D) termination
   (E) adequate

8. CONCENTRATE:
   (A) close eyes
   (B) speak loudly
   (C) pay attention
   (D) look away
   (E) guess randomly

9. Compassionate is to empathetic as
   (A) brilliant is to luminous
   (B) intelligent is to average
   (C) cruel is to kind
   (D) amusing is to dull
   (E) messy is to immaculate

10. Ice skater is to rink as
    (A) ballerina is to stage
    (B) soccer is to goalie
    (C) quiet is to cacophony
    (D) stadium is to hockey player
    (E) child is to park

11. Programmer is to computer as
    (A) library is to encyclopedia
    (B) nurse is to stethoscope
    (C) phone is to receptionist
    (D) coding is to application
    (E) height is to depth

12. If $\sqrt{x}$ is between 9 and 10, which could be the value of x?
    (A) 3
    (B) 5
    (C) 9
    (D) 93
    (E) 109

13. The ratio of boys to girls in a painting club is 3:4. There are 56 girls in the club. How many boys are there?
    (A) 28
    (B) 30
    (C) 32
    (D) 40
    (E) 42

14. Nate watched a movie that was $2\frac{1}{4}$ hours long. He also watched a movie that was 115 minutes long. What was the total length of both movies? *(Note: 1 hour = 60 minutes)*
    (A) 3 hours and 15 minutes
    (B) 3 hours and 30 minutes
    (C) 3 hours and 50 minutes
    (D) 4 hours and 5 minutes
    (E) 4 hours and 10 minutes

15. Peter's pay $30 per week for his farm share. If he wants extra fruits and vegetables, he pays $3 per carton. If Peter paid $48 dollars for his farm share last week, how many extra cartons of fruits and vegetables did he get?
    (A) 4
    (B) 5
    (C) 6
    (D) 7
    (E) 8

16. The coordinate plane represents a town and the grid lines represent streets. What is the shortest distance, in units between the bank and the library, traveling by streets only?

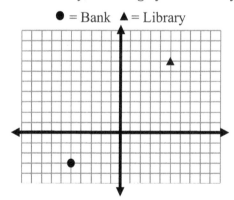

   (A) 10
   (B) 15
   (C) 18
   (D) 20
   (E) 24

STOP
IF YOU FINISH BEFORE TIME IS UP,
CHECK YOUR WORK IN THIS SECTION ONLY.
YOU MAY NOT TURN TO ANY OTHER SECTION.

Scoring the Diagnostic Practice Test (Form A)

Writing Sample – Unscored
Have a parent or trusted educator review the essay or story written for the writing sample. Important areas to focus on include organization, clarity of ideas, originality, and technical precision (spelling, grammar, etc.).

Sections 1-4 – Scored
Score the test using the answer sheet and *referring to the answer key in the back of the book (see table of contents)*.

Step 1: For each section, record the number of questions answered correctly.

Step 2: For each section, record the number of questions answered incorrectly. Then, multiply that number by ¼ to calculate the penalty.

Section	Questions Correct
Quantitative *Section 1 + Section 4*	_____
Reading *Section 2*	_____
Verbal *Section 3*	_____

Section	Questions Incorrect	Penalty
Quantitative *Section 1 + Section 4*	_____	x $\frac{1}{4}$ = _____
Reading *Section 2*	_____	x $\frac{1}{4}$ = _____
Verbal *Section 3*	_____	x $\frac{1}{4}$ = _____

Step 3: For each section, subtract the Penalty in Step 2 from the Questions Correct in Step 1. This is the raw score. Note that the actual test will convert the raw score to a scaled score by comparing the student's performance with all other students in the same grade who took the test.

Section	Raw Score
Quantitative *Section 1 + Section 4*	_____
Reading *Section 2*	_____
Verbal *Section 3*	_____

> **Consider**: How certain were you on the questions you guessed on? Should you have left those questions blank, instead? How should you change the way you guess and leave questions blank?

Carefully consider the results from the diagnostic practice test when forming a study plan. Remember, the Middle Level SSAT is given to students in grades 5-7. Unless the student has finished 7th grade, chances are that there is material on this test that he or she has not yet been taught. If this is the case, and the student would like to improve beyond what is expected of his or her grade, consider working with a tutor or teacher, who can help learn more about new topics.

Section 5 – Unscored
On the real test, the Experimental section will NOT be scored. Consider the student's performance on this section for practice purposes only. Did he or she do better on one section than other? Use this information along with the information from Sections 1-4 to form the study plan.

Quantitative

Overview

The Quantitative sections assess a student's command over various mathematics topics, including algebra, pre-algebra, geometry, probability, statistics, and number theory.

There are two Quantitative sections on the Middle Level SSAT, both of which are scored.

On the Actual Test

Each of the two quantitative sections contain 25 questions (for a total of 50 math questions on the entire scored test).

Students have 30 minutes each to complete the two sections (for a total of 60 minutes on the entire scored test).

Every question in the Quantitative sections is multiple choice. There will be one question followed by five answer choices (A through E). Students are given blank space to the right of each question where they can do their work.

In This Practice Book

Below are the main content areas that are included in the Middle Level SSAT. A list of subtopics can be found in the table of contents.

- Numbers Concepts & Operations
- Pre-Algebra
- Algebra
- Geometry
- Measurements
- Data Analysis
- Statistics & Probability

Considering the results of your diagnostic practice test, we recommend that students focus on the topics that are most challenging to them. Since there may be material in this workbook that they have not yet learned in school, we also encourage students to seek additional help from a trusted teacher or tutor to enhance their knowledge of those subjects.

Treat each question as you would questions on the real test. This means practicing whether or not to answer the question or leave it unanswered. **Remember**, on the actual test, answering a question correctly earns you 1 point, but answering a question incorrectly means you lose ¼ of a point. Therefore, attempt every question for the sake of practice (and read the online answer explanations), but try practicing leaving questions unanswered. Then, see whether or not you would have answered the question correctly.

The questions in each section are progressive, which means they start out easier, and become more and more difficult as they build on the concepts related to that topic. If students find that some questions are tricky, they should consider asking an educator for help. Don't be discouraged!

Tutorverse Tips!

You won't be able to use a calculator on the test. If, as you are answering a question, things start to get more and more complicated, take a step back and think about what the question is asking you to do. If necessary,

use the answer choices themselves to help you arrive at the correct answer by plugging them into formulas or expressions.

You do **not** have to memorize customary unit conversion tables (for instance, the number of feet in a mile), as any such information will be provided. However, metric unit conversions will **not** be provided (i.e. the number of milliliters in a liter).

Guessing

Knowing whether or not to guess can be tricky. The Middle Level SSAT gives you 1 point for each question that is answered correctly. However, if you answer a question incorrectly, ¼ of a point will be deducted from your total score (see example below).

The formula for determining the raw score is:
(Number of Questions Answered Correctly × 1) – (Number of Questions Answered Incorrectly × ¼)

No points will be awarded or deducted for questions that are left unanswered. Therefore, answer easy questions first, and come back to tougher questions later.

<u>How Guessing Impacts A Score</u>

Since there are 150 scored questions (167 total questions – 1 writing prompt – 16 experimental questions), the highest possible raw score is 150 points. This would be awarded to students who answer all 150 questions correctly.

If a student answers 110 questions correctly **but answers 40 questions incorrectly**, he or she will have earned (110 × 1) – (40 × ¼), or 110 – 10 = 100 points.

If a student answers 110 questions correctly **but leaves 40 questions unanswered**, he or she will have earned (110 × 1) – (40 × 0), or 110 – 0 = 110 points.

Therefore, it is important to refrain from making wild guesses. Instead, try to use process of elimination to make an educated guess.

Number Concepts & Operations

Place Value

1. What is the value of the underlined digit in 9,247.86?
   (A) nine thousandths
   (B) nine hundredths
   (C) nine
   (D) nine hundred
   (E) nine thousand

2. Which number has a value that is exactly $\frac{1}{100}$ of the value 82,159?
   (A) 8,215.9
   (B) 821.59
   (C) 82.159
   (D) 8.2159
   (E) 0.82159

3. The ■ in the number series below represents a digit from 0 through 9. If the number is less than 7,264, what is the greatest possible value for ■?
   7, 2, ■, 4
   (A) 0
   (B) 2
   (C) 4
   (D) 5
   (E) 6

4. Of the following lists, which will contain equal numbers when each number in the list is rounded to the nearest tenth?
   (A) 2.45, 2.46, 2.3, 2.33
   (B) 3.21, 3.19, 3.25, 3.15
   (C) 5.79, 5.82, 5.76, 5.78
   (D) 8.01, 8.03, 8.04, 8.05
   (E) 9.61, 9.54, 9.62, 9.64

5. In the number 927.46, which of the following digits is in the ones place?
   (A) 2
   (B) 4
   (C) 6
   (D) 7
   (E) 9

6. Which value is equal to 47.901 × 0.001?
   (A) 47.901
   (B) 479.01
   (C) 4.7901
   (D) 0.47901
   (E) 0.047901

7. Which expression is equivalent to 4.95×10^3?
   (A) 0.495×10^4
   (B) 49.5×10^3
   (C) $4,950 \times 10$
   (D) 0.0495×10^4
   (E) 49.5×10^4

8. Which expression has a value equivalent to the value of the digit 5 in 6.051?
   (A) 0.05 × 10
   (B) 5 × 0.1
   (C) 0.5 × 0.1
   (D) 50 × 0.01
   (E) 0.5 × 10

9. What is the value of the underlined digit in 4,820.175?
   (A) seven thousandths
   (B) seven hundredths
   (C) seven tenths
   (D) seven
   (E) seven hundred

10. Which number has a value of 6,000 when it is multiplied by 0.1?
    (A) 0.6
    (B) 6
    (C) 60
    (D) 600
    (E) 60,000

11. Which expression has the same value as the value of the underlined digit in 459.21?
    (A) 9 × 0.1
    (B) 0.9 × 0.1
    (C) 9 ÷ 10
    (D) 0.9 × 10
    (E) 90 ÷ 100

Decimals

1. What is 4.2 + 3.5 − 0.8?
   (A) −0.3
   (B) −0.1
   (C) 1.5
   (D) 6.9
   (E) 8.7

2. What is 1.2 − 5.3 + 8.8?
   (A) −12.9
   (B) 2.3
   (C) 3.5
   (D) 4.7
   (E) 15.3

3. What is 6.4 × 2.5?
   (A) 0.16
   (B) 1.6
   (C) 16
   (D) 160
   (E) 1600

4. What is 2.8 × 12.5?
   (A) 3.5
   (B) 35
   (C) 153
   (D) 350
   (E) 3,500

5. Compute the product of (1.2)(0.25).
   (A) 300
   (B) 30
   (C) 6
   (D) 3
   (E) 0.3

6. Find the quotient when 50.4 is divided by 0.042.
   (A) 0.12
   (B) 1.2
   (C) 12
   (D) 120
   (E) 1,200

7. Find the value of $(0.5)^3$.
   (A) 125
   (B) 12.5
   (C) 1.25
   (D) 0.125
   (E) 0.000125

8. Which statement is false?
   (A) (0.3)(0.5) < 0.2
   (B) (0.3)(0.6) > 0.2
   (C) (0.3)(0.2) > 0.05
   (D) (0.5)(0.6) > 0.2
   (E) (0.2)(0.6) > 0.1

9. Find the value of $\sqrt{0.49}$.
   (A) 0.0007
   (B) 0.007
   (C) 0.07
   (D) 0.7
   (E) 7

10. Find the quotient when 0.525 is divided by 3.5.
    (A) 150
    (B) 15
    (C) 1.5
    (D) 0.15
    (E) 0.015

11. Compute the product (0.28)(28).
    (A) 784
    (B) 78.4
    (C) 7.84
    (D) 0.784
    (E) 0.0784

12. Find the difference 67.5 − 0.0576.
    (A) 67.4424
    (B) 66.924
    (C) 66.24
    (D) 61.74
    (E) 67.4534

Fractions

1. Which of the following is equivalent to $\frac{2}{3}$?
   - (A) $\frac{3}{6}$
   - (B) $\frac{3}{9}$
   - (C) $\frac{6}{10}$
   - (D) $\frac{12}{15}$
   - (E) $\frac{12}{18}$

2. Reduce $\frac{12}{72}$ to lowest terms.
   - (A) $\frac{1}{8}$
   - (B) $\frac{1}{6}$
   - (C) $\frac{1}{5}$
   - (D) $\frac{1}{4}$
   - (E) $\frac{2}{3}$

3. Which of the following is the sum of $\frac{1}{5}$ and $\frac{3}{4}$?
   - (A) $\frac{1}{20}$
   - (B) $\frac{4}{9}$
   - (C) 1
   - (D) $\frac{17}{20}$
   - (E) $\frac{19}{20}$

4. Which of the following is the sum of $\frac{1}{3}$ and $\frac{5}{6}$?
   - (A) $\frac{7}{12}$
   - (B) $\frac{2}{3}$
   - (C) 1
   - (D) $1\frac{1}{6}$
   - (E) $1\frac{1}{3}$

5. What is the value of $\frac{1}{3}(-\frac{1}{2} \times \frac{2}{7})$?
   - (A) $-\frac{7}{12}$
   - (B) $-\frac{3}{14}$
   - (C) $-\frac{1}{21}$
   - (D) $\frac{1}{21}$
   - (E) $\frac{7}{12}$

6. What is the value of $\frac{1}{8}(\frac{3}{5} \times \frac{1}{3})$?
   - (A) $\frac{1}{40}$
   - (B) $\frac{1}{8}$
   - (C) $\frac{3}{8}$
   - (D) $\frac{1}{2}$
   - (E) $\frac{1}{4}$

7. What is $\frac{1}{6} - \frac{1}{5}$?
   - (A) $-\frac{1}{30}$
   - (B) $-\frac{2}{15}$
   - (C) $-\frac{1}{10}$
   - (D) 1
   - (E) $\frac{1}{30}$

8. The shape shown is divided into 5 congruent squares. What fraction of the shape is shaded?

   - (A) $\frac{1}{10}$
   - (B) $\frac{1}{5}$
   - (C) $\frac{2}{5}$
   - (D) $\frac{3}{10}$
   - (E) $\frac{3}{5}$

9. Samantha is painting a wall. She painted $\frac{1}{2}$ of the wall yesterday, and $\frac{1}{4}$ of the remaining wall today. If the area of the wall is 120 ft², how much of the wall does Samantha have left to paint?
   - (A) 15 ft²
   - (B) 45 ft²
   - (C) 60 ft²
   - (D) 75 ft²
   - (E) 100 ft²

10. What is the value of $\frac{1}{4} \times \frac{1}{2} \times \frac{2}{3}$?
    (A) $\frac{4}{9}$
    (B) $\frac{1}{6}$
    (C) $\frac{1}{7}$
    (D) $\frac{1}{9}$
    (E) $\frac{1}{12}$

11. Which of the following expresses the list below in order from least to greatest?
    $87\%, \frac{1}{5}, 0.21, \frac{1}{2}, 19\%, 0.45$
    (A) $19\%, 0.21, \frac{1}{2}, \frac{1}{5}, 0.45, 87\%$
    (B) $\frac{1}{2}, \frac{1}{5}, 19\%, 0.21, 0.45, 87\%$
    (C) $19\%, \frac{1}{5}, 0.21, 0.45, \frac{1}{2}, 87\%$
    (D) $0.21, 0.45, 19\%, \frac{1}{2}, \frac{1}{5}, 87\%$
    (E) $\frac{1}{5}, 0.21, 19\%, 0.45, 87\%, \frac{1}{2}$

12. How many fifths are there in 3.4?
    (A) 2
    (B) 4
    (C) 15
    (D) 17
    (E) 20

13. If $\frac{1}{2} = \frac{5}{y}$, what is the value of y?
    (A) 3
    (B) 5
    (C) 10
    (D) 15
    (E) 20

14. If $\frac{4}{7} = \frac{x}{21}$, what is the value of x?
    (A) 8
    (B) 10
    (C) 11
    (D) 12
    (E) 28

15. Which is equivalent to $4\frac{1}{3} + 2\frac{2}{9}$?
    (A) $4\frac{5}{9}$
    (B) $6\frac{1}{4}$
    (C) $6\frac{5}{9}$
    (D) $7\frac{4}{9}$
    (E) $7\frac{5}{9}$

16. What is $\frac{15}{8} \div 5$?
    (A) $-\frac{25}{8}$
    (B) $\frac{3}{8}$
    (C) $\frac{5}{4}$
    (D) $\frac{5}{2}$
    (E) $\frac{15}{8}$

17. What is $\frac{21}{4} \div \frac{14}{3}$?
    (A) $\frac{49}{2}$
    (B) 7
    (C) 5
    (D) $\frac{9}{8}$
    (E) $\frac{8}{9}$

Percents

1. What is 18% of 25?
   (A) 3.7
   (B) 4.5
   (C) 5
   (D) 9
   (E) 18

2. Twenty-four percent of seventy-five is x. What is the value of x?
   (A) 8
   (B) 15
   (C) 18
   (D) 57
   (E) 1,800

3. If 20% of a number is 80, what is 5% of the number?
   (A) 0.8
   (B) 4
   (C) 16
   (D) 20
   (E) 40

4. 8 is 40% of what number?
   (A) 32
   (B) 20
   (C) 16
   (D) 10.5
   (E) 3.2

5. If 40 is 25% of x, what is the value of x?
   (A) 10
   (B) 16
   (C) 80
   (D) 100
   (E) 160

6. If x is 20% of 50, what is x% of 50?
   (A) 2
   (B) 5
   (C) 10
   (D) 20
   (E) 25

7. 120 is what percent of 80?
   (A) 50
   (B) $66\frac{2}{3}$
   (C) $41\frac{2}{3}$
   (D) 150
   (E) 200

8. 60 is 125% of what number?
   (A) 15
   (B) 48
   (C) 120
   (D) 240
   (E) 300

9. The Mountain Street Middle School Student Council has an annual budget of $4,000. The circle graph below displays how they spent the $4,000 last year.

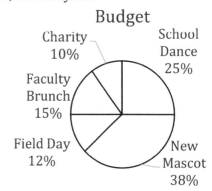

   On which expense did they spend $1,000?
   (A) Field Day
   (B) New Mascot Costume
   (C) Faculty Brunch
   (D) Charity
   (E) School Dance

10. Kai surveyed 50 people to find out people's favorite hot dog toppings. He made this chart:

Topping	Number of People
Ketchup	22
Mustard	18
Relish	3
None	7

   What percent of people chose Mustard or Ketchup?
   (A) 20
   (B) 36
   (C) 40
   (D) 44
   (E) 80

11. 30 is what percent of 45?
    (A) 30
    (B) 50
    (C) $66\frac{2}{3}$
    (D) 75
    (E) 150

12. The number 80 is increased by 75% and the result is decreased by 20% to give the number *n*. What is the value of *n*?
    (A) 12
    (B) 28
    (C) 48
    (D) 112
    (E) 124

13. In Joe's third grade class 30% of the students are left-handed. If there are 6 left-handed students in Joe's class, how many total students are in the class?
    (A) 12
    (B) 15
    (C) 18
    (D) 20
    (E) 24

14. Julius made a reservation at a restaurant for a party. After the party, he received a bill for $300 after tax. If Julius wants to leave a 15% tip, how much will he have to pay?
    (A) 315
    (B) 330
    (C) 345
    (D) 360
    (E) 375

15. What percent of the boxes have either letters or odd numbers?

1	2	3		
4	5	6		
7	8	9		
		A	B	C
		D	E	F

    (A) 10%
    (B) 11%
    (C) 40%
    (D) 44%
    (E) 60%

16. Bradley Middle School held a dance for their seventh and eighth grade students. 60% of the students in attendance were seventh graders and 40% were eighth graders. If 90 seventh graders attended, how many total students attended the dance?
    (A) 225
    (B) 150
    (C) 135
    (D) 60
    (E) 54

17. The Jackette Shoppe had a weekend sale. All coats and jackets were on sale for 40% off. On Monday, the price of anything left in the store increased 40%. If a jacket cost $150 before the sale weekend and did not sell over the weekend, what was its new price on Monday?
    (A) $150
    (B) $126
    (C) $90
    (D) $54
    (E) $24

Decimals/Fractions/Percents

1. Which fraction is equivalent to 9%?
    (A) $\frac{1}{9}$
    (B) $\frac{1}{90}$
    (C) $\frac{9}{100}$
    (D) $\frac{9}{90}$
    (E) $\frac{9}{900}$

2. What is the value of $\frac{2}{4} \div 0.8$?
    (A) $\frac{2}{5}$
    (B) $\frac{5}{8}$
    (C) $\frac{1}{4}$
    (D) $\frac{3}{4}$
    (E) $\frac{3}{10}$

3. Which decimal has a value between $\frac{1}{3}$ and $\frac{2}{3}$?
   (A) 0.75
   (B) 0.69
   (C) 0.44
   (D) 0.30
   (E) 0.25

4. What is 20% of $\frac{1}{10}$?
   (A) $\frac{1}{5}$
   (B) $\frac{1}{20}$
   (C) $\frac{1}{50}$
   (D) $\frac{1}{100}$
   (E) $\frac{1}{200}$

5. Which pair of values is *not* equivalent?
   (A) $\frac{1}{4}$ and 25%
   (B) $\frac{3}{5}$ and 60%
   (C) $\frac{4}{8}$ and 50%
   (D) $\frac{7}{10}$ and 75%
   (E) $\frac{80}{100}$ and 80%

6. Which fraction has a value greater than 20% and less than 45%?
   (A) $\frac{1}{2}$
   (B) $\frac{1}{5}$
   (C) $\frac{2}{3}$
   (D) $\frac{3}{5}$
   (E) $\frac{1}{3}$

7. The set of numbers in order from least to greatest is
   (A) 0.2, $\frac{2}{5}$, 0.45, $\frac{5}{6}$
   (B) 0.2, 0.45, $\frac{2}{5}$, $\frac{5}{6}$
   (C) $\frac{2}{5}$, 0.2, 0.45, $\frac{5}{6}$
   (D) 0.2, $\frac{5}{6}$, 0.45, $\frac{2}{5}$
   (E) $\frac{2}{5}$, $\frac{5}{6}$, 0.2, 0.45

8. Ashley's dinner bill was paid for evenly by 5 people. What percent of Ashley's bill did each person pay?
   (A) 20%
   (B) 40%
   (C) 50%
   (D) 60%
   (E) 75%

9. $\frac{8}{11} \times 2.2 \div 0.8 =$
   (A) 128
   (B) 5
   (C) $\frac{41}{11}$
   (D) $\frac{38}{11}$
   (E) 2

10. $1.5 \div \frac{9}{20} \times 1.2 =$
    (A) 0.01
    (B) 0.63
    (C) 0.81
    (D) 4
    (E) 324

11. Which fraction is equivalent to 0.35?
    (A) $\frac{1}{35}$
    (B) $\frac{7}{100}$
    (C) $\frac{1}{5}$
    (D) $\frac{7}{20}$
    (E) $\frac{35}{10}$

12. The set of numbers in order from least to greatest is
    (A) $\frac{1}{2}$, 0.1, $\frac{2}{3}$, 0.23
    (B) 0.1, 0.23, $\frac{1}{2}$, $\frac{2}{3}$
    (C) $\frac{1}{2}$, $\frac{2}{3}$, 0.1, 0.23
    (D) 0.23, $\frac{2}{3}$, 0.1, $\frac{1}{2}$
    (E) $\frac{2}{3}$, $\frac{1}{2}$, 0.23, 0.1

13. Which percent has a value between $\frac{1}{2}$ and $\frac{3}{5}$?
    (A) 25%
    (B) 43%
    (C) 57%
    (D) 66%
    (E) 75%

14. Which pair of values is not equivalent?
    (A) $\frac{1}{10}$ and 0.1
    (B) $\frac{1}{5}$ and 0.2
    (C) $\frac{1}{4}$ and 0.25
    (D) $\frac{1}{3}$ and 0.3
    (E) $\frac{1}{2}$ and 0.5

15. Eli and five of his friends went to a football game. They all agreed to split the cost evenly. What percent of the cost did each of Eli and his friends agree to pay?
    (A) 6%
    (B) 12%
    (C) 15%
    (D) $16\frac{2}{3}$%
    (E) 20%

16. $\frac{3}{4} \times 8 \div 0.2 =$
    (A) 3
    (B) 6
    (C) 12
    (D) 30
    (E) 48

Whole Numbers

1. What number is halfway between 37 and 49?
   (A) 40
   (B) 43
   (C) 44.5
   (D) 46
   (E) 47

2. Which number is evenly divisible by 6?
   (A) 117
   (B) 124
   (C) 130
   (D) 138
   (E) 145

3. Of the following, which is closest to the value of $\frac{99 \times 100^2}{8 \times 13}$?
   (A) 500
   (B) 1,000
   (C) 5,000
   (D) 10,000
   (E) 100,000

4. Which is the closest value of $\frac{48 \times 3^2}{98 \times 101}$?
   (A) $\frac{1}{10}$
   (B) $\frac{1}{20}$
   (C) $\frac{1}{100}$
   (D) $\frac{1}{200}$
   (E) $\frac{1}{1,000}$

5. Which of the following lists contains only prime numbers?
   (A) 1, 3, 5, 7, 11
   (B) 2, 4, 6, 8, 10
   (C) 2, 5, 7, 11, 17
   (D) 5, 11, 17, 19, 21
   (E) 13, 15, 21, 23, 25

6. Which of the following lists contains only prime numbers?
   (A) 4, 7, 9, 12, 13
   (B) 20, 23, 29, 30, 31
   (C) 37, 41, 43, 47, 53
   (D) 53, 55, 59, 61, 63
   (E) 71, 72, 74, 83, 87

7. If $\frac{x}{30}$ is between 7 and 8, then x can equal
   - (A) 23
   - (B) 189
   - (C) 220
   - (D) 256
   - (E) 271

8. If $\frac{n}{9}$ is a whole number, which of the following could be a value of n?
   - (A) 83
   - (B) 117
   - (C) 136
   - (D) 188
   - (E) 244

9. If $21 + 21 + 21 + 21 = 4 \times k$, what is the value of k?
   - (A) 4
   - (B) 21
   - (C) 25
   - (D) 42
   - (E) 84

10. The sum of three consecutive integers is 138. What is the value of the greatest integer?
    - (A) 44
    - (B) 45
    - (C) 46
    - (D) 47
    - (E) 48

11. What is the greatest common factor of 64 and 120?
    - (A) 6
    - (B) 8
    - (C) 12
    - (D) 64
    - (E) 960

12. When 69 is divided by p, the remainder is 3. Which could be the value of p?
    - (A) 3
    - (B) 11
    - (C) 15
    - (D) 26
    - (E) 72

Order of Operations

1. Calculate: $3(10 - 8) + 6$
   - (A) 6
   - (B) 12
   - (C) 20
   - (D) 24
   - (E) 28

2. What is the value of $100 - 15 + 20 \div 2$?
   - (A) 52.5
   - (B) 75
   - (C) 95
   - (D) 105
   - (E) 125

3. What is the value of 0.5^3?
   - (A) 12.5
   - (B) 1.5
   - (C) 1.25
   - (D) 0.15
   - (E) 0.125

4. Which is the value of $(9 - 7)^3 - 10$?
   - (A) –6
   - (B) –2
   - (C) 0
   - (D) 2
   - (E) 10

5. Find the value of $2 + 4 \times 3^2 \div 12 + 4^2$
   - (A) 20.5
   - (B) 21
   - (C) 22
   - (D) 43
   - (E) 961

6. Calculate: $7 + 14 \div 2 \times 3^2 - 8$
   - (A) 72
   - (B) 62
   - (C) 14
   - (D) 7
   - (E) –2

7. Find the value of the expression:
   $7 + 6 \times (8 - 9)^3 \div 2$
   (A) 0.5
   (B) 4
   (C) 6.5
   (D) 7
   (E) 10

8. What is the product of $10 - 5 \times 9$ and $17 - 2^4$?
   (A) 46
   (B) 35
   (C) 19
   (D) −34
   (E) −35

9. What is the value of $-2[(5 + 3)^2 - 9 \times 4]$?
   (A) 28
   (B) −40
   (C) −50
   (D) −56
   (E) −440

10. What is the value of $5[16 \div 2 + (12 - 10)^2]$
    (A) 12
    (B) 17
    (C) 44
    (D) 50
    (E) 60

11. What is the value of the expression, $46 + 3(x + y)$, if $x = 7$ and $y = 4$?
    (A) 13
    (B) 25
    (C) 54
    (D) 63
    (E) 79

12. Which is the value of the expression $(8 \div 2)^2 \div (5 - 3)^2 \times 2$?
    (A) 1
    (B) 2
    (C) 4
    (D) 8
    (E) 128

Pre-Algebra

Ratio and Proportions

1. A cat chased away 3 out of every 4 mice in a house. If 36 mice were chased away, how many mice remain?
   (A) 9
   (B) 12
   (C) 16
   (D) 18
   (E) 48

2. Two squirrels can eat 8 acorns in 4 days. How many days would it take four squirrels to eat 72 acorns?
   (A) 6
   (B) 9
   (C) 12
   (D) 18
   (E) 36

3. A perfect test score on Mrs. Bates' test is 100 points. The test has 50 multiple choice questions equally weighted. David scored 86 points. How many questions did he answer correctly?
   (A) 86
   (B) 46
   (C) 44
   (D) 43
   (E) 42

4. For every two cups of water Robert drank, Mike drank three cups of water. Mike drank 144 cups of water in a month. How many did Robert drink that month?
   (A) 48
   (B) 72
   (C) 96
   (D) 108
   (E) 216

5. A lion ate 5 pounds of meat per day. A tiger ate 3 pounds of meat per day. Together, in a certain number of days, they ate 168 pounds of meat. How many pounds did the tiger eat?
   (A) 21
   (B) 63
   (C) 84
   (D) 105
   (E) 231

6. 80 dogs are in a kennel; all are either black or brown. The ratio of black dogs to brown dogs in the kennel is 2:3. How many brown dogs are in the kennel?
   (A) 24
   (B) 32
   (C) 40
   (D) 48
   (E) 72

7. There were 40 adults and 24 children at the art museum. Which statement is NOT true?
   (A) The ratio of adults to children was 5:3.
   (B) The ratio of children to adults was 6:10.
   (C) 62.5% of the people were adults.
   (D) The ratio of adults to total people is 1:2.
   (E) The ratio of children to total people is 6:16.

8. If you can buy 5 tacos for as much money as 2 enchiladas, and a taco costs $4.80, how much does an enchilada cost?
   (A) $1.92
   (B) $2.40
   (C) $6.00
   (D) $9.60
   (E) $12.00

9. A tractor can plow 18,000 square feet in a day. At this rate, how many square feet can a tractor plow in $\frac{1}{12}$ of a day?
   (A) 150
   (B) 300
   (C) 1,500
   (D) 3,000
   (E) 216,000

10. What is 0.68 expressed as a fraction with a denominator of 25?
    (A) $\frac{16}{25}$
    (B) $\frac{17}{25}$
    (C) $\frac{18}{25}$
    (D) $\frac{19}{25}$
    (E) $\frac{21}{25}$

11. On a state test, for every problem Ben missed, he got 7 others right. What percent did he get right?
    (A) 14
    (B) 80
    (C) 87.5
    (D) 90
    (E) 95

12. For every 12 students who agreed to take a survey, 15 others declined. 108 students were asked to take it. How many declined?
    (A) 48
    (B) 54
    (C) 60
    (D) 72
    (E) 90

13. In a group of 48 cats, for every cat that is awake, 5 were sleeping. How many cats are sleeping?
    (A) 8
    (B) 12
    (C) 26
    (D) 32
    (E) 40

14. A certain jaguar was successful in $\frac{2}{5}$ of its attempts to hunt deer. 60 deer escaped from this jaguar. How many deer did the jaguar attempt to hunt in total?
    (A) 24
    (B) 36
    (C) 90
    (D) 96
    (E) 100

15. Carla won 3 out of every 7 chess matches she played. If she won 21 matches, how many did she play?
    (A) 9
    (B) 11
    (C) 28
    (D) 35
    (E) 49

16. Kevin paints two walls in one day. Brian paints three walls in one day. Together, over several days, they painted 95 walls. How many walls to Kevin paint?
    (A) 19
    (B) 38
    (C) 57
    (D) 76
    (E) 190

17. The convenience store is having a deal on candy bars where if you buy 3 candy bars you get the 4th for free. If you but 12 candy bars, how many will you get for free?
    (A) 1
    (B) 3
    (C) 4
    (D) 9
    (E) 16

18. The ratio of dogs to cats to gerbils at a pet store is 2:3:1. If there are 16 dogs at the pet store, how many gerbils are there?
    (A) 4
    (B) 8
    (C) 16
    (D) 24
    (E) 32

19. A perfect total score from five judges at a gymnastics tournament is a 75. If all the judges can give the same range of scores, what is the highest score a single judge can give?
    (A) 5
    (B) 10
    (C) 15
    (D) 25
    (E) 30

20. A truck driver can drive 800 miles in a day. How many miles can that truck driver drive in $\frac{1}{5}$ of a day?
    (A) 100
    (B) 160
    (C) 200
    (D) 400
    (E) 4,000

21. On his most recent science quiz, Spenser got 2 out of every 9 questions incorrect. If there were 36 questions on the quiz, how many did Spenser answer correctly?
    (A) 8
    (B) 16
    (C) 28
    (D) 30
    (E) 36

22. What is 0.28 expressed as a fraction with the denominator of 25?
    (A) $\frac{6}{25}$
    (B) $\frac{7}{25}$
    (C) $\frac{8}{25}$
    (D) $\frac{9}{25}$
    (E) $\frac{11}{25}$

23. For every 3 students that completed the extra-credit assignment, 5 other students did not. If 48 students were offered the extra-credit assignment, how many did finish the assignment?
    (A) 12
    (B) 18
    (C) 24
    (D) 30
    (E) 36

24. Leo was successful in $\frac{5}{7}$ of his field goal attempts. If he missed 10 fields goals, how many did he attempt?
    (A) 8
    (B) 14
    (C) 20
    (D) 25
    (E) 35

Sequences, Patterns and Logic

1. The series of shapes below repeats indefinitely, as shown. What is the 13th shape in the sequence?

   (A) ★
   (B) ☾
   (C) △
   (D) ○
   (E) ✚

2. If *n* is a positive even integer, which of the following must be an even integer?
    (A) $2n + 1$
    (B) $3n + 1$
    (C) $2n - 1$
    (D) $n + 3$
    (E) $\frac{n \times 4}{2}$

3. If *n* is an odd integer, which of the following must be an integer?
    (A) $\frac{n}{2}$
    (B) $\frac{n}{3}$
    (C) $\frac{3(n+1)}{2}$
    (D) $\frac{3n}{2}$
    (E) $\frac{(n+2)}{4}$

4. What is the next number in the series below?
   2, 5, 11, 23, ___
    (A) 29
    (B) 35
    (C) 41
    (D) 47
    (E) 49

5. What is the next number in the series?
   8, 12, 18, 26, ___
    (A) 34
    (B) 36
    (C) 38
    (D) 40
    (E) 42

6. Teddy is stringing beads together in the following pattern: one red, one pink, one orange, one purple, and one green. What color will the 87th term be?
    (A) red
    (B) pink
    (C) orange
    (D) purple
    (E) green

7. The first number in a series is 3. The second number is 4. Each number after the second is the sum of the two preceding numbers. What is the 7th number in the series?
    (A) 18
    (B) 29
    (C) 37
    (D) 46
    (E) 47

8. What is the next number in the series?

    $6\frac{1}{2}, 7\frac{1}{2}, 7, 8, 7\frac{1}{2}, 8\frac{1}{2}, 8,$ ___

    (A) 7
    (B) $7\frac{1}{2}$
    (C) 8
    (D) $8\frac{1}{2}$
    (E) 9

9. In the series below, $\frac{2}{5}$, is the first term. For each term after the first, the numerator and the denominator are each two more than the numerator and the denominator in the preceding term. What is the value of x?

    $\frac{2}{5}, \frac{4}{7}, \frac{6}{9}, \frac{8}{11}, \frac{10}{13}, ..., \frac{22}{x}$

    (A) 15
    (B) 17
    (C) 19
    (D) 23
    (E) 25

10. The first seven shapes of a pattern are shown.

    What is the 11th term?

    (A)
    (B)
    (C)
    (D)
    (E)

11. The series below repeats indefinitely.

    What is the 29th term in the series?

    (A) ☆
    (B) ☾
    (C) △
    (D) ○
    (E) □

12. Based on the equations below, what is the value of □?

    ○ = 20
    ○ × △ = □
    △ + ○ = 45

    (A) 25
    (B) 45
    (C) 400
    (D) 450
    (E) 500

13. If x is a negative integer, which of the following must be a positive even integer?

    (A) $x + □$
    (B) $x + 2$
    (C) $2x$
    (D) $-2x$
    (E) $-3x + 1$

Estimation

1. One pen costs $0.48. Of the following, which is the closest to the price of 4 of these pens?
    (A) $1.00
    (B) $1.50
    (C) $2.00
    (D) $2.50
    (E) $3.00

2. Of the following sums, which has a value that is closest to 32 + 19 + 38?
    (A) 40 + 20 + 40
    (B) 30 + 20 + 40
    (C) 30 + 20 + 30
    (D) 30 + 10 + 30
    (E) 20 + 10 + 30

3. Of the following, which is the closest to 89.5 – 11.8?
   (A) 60
   (B) 65
   (C) 72
   (D) 78
   (E) 83

4. Shelby spent $123.05 on a coat, $228.54 on clothes, and $113.24 on shoes. Which is the best estimate for the total cost of Shelby's purchases when rounding each amount to the nearest $10?
   (A) $380
   (B) $400
   (C) $460
   (D) $480
   (E) $500

5. Michael's house is 5.88 miles from school. 1 mile is equivalent to 5,280 feet. Which is the most reasonable approximate length, in feet, of the distance between Michael's house and school?
   (A) 3,000
   (B) 5,000
   (C) 6,000
   (D) 30,000
   (E) 300,000

6. A publisher shipped 208 boxes of books to bookstores. There are 48 books in each box. Approximately how many books did the publisher ship?
   (A) 12,000
   (B) 10,000
   (C) 8,000
   (D) 6,000
   (E) 1,000

7. A truck delivered 1,764 cases of water to a store. Approximately how many bottles of water are there if each case holds 12 bottles?
   (A) 200,000
   (B) 180,000
   (C) 18,000
   (D) 10,000
   (E) 2,000

8. A school bus can hold up to 55 students. If each bus is filled, approximately how many buses are needed for 1,168 students?
   (A) 10
   (B) 20
   (C) 40
   (D) 80
   (E) 200

9. A recipe calls for 18 ounces of flour and 11 ounces of sugar. Approximately how many ounces of flour and sugar combined are needed to make 28 recipes?
   (A) 90
   (B) 120
   (C) 600
   (D) 900
   (E) 6,000

10. A secretary can type 98 words per minute. Approximately how many words can she type in 123 minutes?
    (A) 1.2
    (B) 300
    (C) 1,200
    (D) 3,000
    (E) 12,000

11. Andrew went out to dinner with eight of his friends. Their bill came to a total of $216.38. If they split the bill evenly, approximately how much will Andrew have to pay?
    (A) $18
    (B) $22
    (C) $27
    (D) $28
    (E) $31

Algebra

Interpreting Variables

1. If $a = 4$ and $b = 6$, then $3ab$ equals
   - (A) 13
   - (B) 27
   - (C) 30
   - (D) 72
   - (E) 346

2. If c is 8 less than d, then d must equal
   - (A) 8 more than c
   - (B) 8 minus c
   - (C) 8 divided by c
   - (D) 8 times c
   - (E) $\frac{1}{8}$ of c

3. At a certain store, a jar of peanut butter costs $6 and a jar of jelly costs $4. If the total cost of the jars that Kelly bought can be represented by $6x + 4y$, what does y represent?
   - (A) the cost of a jar of peanut butter
   - (B) the cost of a jar of jelly
   - (C) the total cost of the items
   - (D) the number of jars of peanut butter bought
   - (E) the number of jars of jelly bought

4. The length of one side of a square is $5a - 2$. What is the perimeter of the square?
   - (A) $20a$
   - (B) $20a - 2$
   - (C) $20a - 8$
   - (D) $10a - 4$
   - (E) $5a^2$

5. Ting bought g cartons of eggs. Each carton contains 12 eggs. Which expression best represents how many eggs Ting bought?
   - (A) $\frac{g}{12}$
   - (B) $12 + g$
   - (C) $12 - g$
   - (D) $12g$
   - (E) $\frac{12}{g}$

6. The width of a rectangle is $(b + 5)$ and its length is 3 times larger than its width. What is the perimeter of this rectangle?
   - (A) $3b + 15$
   - (B) $5b + 10$
   - (C) $6b + 15$
   - (D) $5b + 20$
   - (E) $5b + 40$

7. Jennifer's age is represented by a. Henry is 5 years older than Jennifer. Todd is 9 years younger than Henry. Which expression represents Todd's age in terms of a?
   - (A) $a - 14$
   - (B) $a - 4$
   - (C) $a + 4$
   - (D) $a + 9$
   - (E) $a + 13$

8. Callie has $500 to spend on tables and chairs for his patio. A table costs $150, and a chair costs $70. If Callie will purchase one table, which inequality best represents how many chairs, x, she can buy?
   - (A) $70x \div 150 \leq 500$
   - (B) $70x - 150 \geq 500$
   - (C) $70x - 150 \leq 500$
   - (D) $70x + 150 \geq 500$
   - (E) $70x + 150 \leq 500$

9. At a food truck, a hot dog costs $4 and an order of fries costs $3. If the total cost of the order that Gus made can be represented by $4x + 3y$, what does x represent?
   - (A) the cost of a hot dog
   - (B) the cost of an order of fries
   - (C) the total cost of the items
   - (D) the number of hot dogs in the order
   - (E) the number of orders of fries in the order

10. At a clothing store, customers must spend at least $75 in order to receive a 20% discount on their entire purchase. Rhonda currently has $62 worth of items in her basket. Which inequality, in d dollars, represents how much more she needs to spend in order to receive the discount?
    (A) $d \geq 13$
    (B) $d < 13$
    (C) $d \geq 75$
    (D) $d > 20$
    (E) $d \leq 62$

11. Matthew has $100 to spend on seeds and dirt for his garden. A packet of seeds costs $2 each, and a bag of dirt costs $10. If Matthew will purchase one bag of dirt, which inequality best represents how many packets of seeds, x, he can buy?
    (A) $2x \div 10 \leq 100$
    (B) $2x - 10 \geq 100$
    (C) $2x - 10 \leq 100$
    (D) $2x + 10 \geq 100$
    (E) $2x + 10 \leq 100$

12. Tyrone has 6 hangers in each of his closets at home. He has at least 24 hangers in total. Which inequality best represents how many closets he has in his home?
    (A) $x \geq 24$
    (B) $6x \geq 24$
    (C) $6x \leq 24$
    (D) $6 \geq 24x$
    (E) $6 \leq 24x$

13. Sally's height, in inches, is represented by h. Jimmy is 5 inches taller than Sally. Betty is 3 inches shorter than Jimmy. Which expression represents Betty's height in terms of h?
    (A) $h - 2$
    (B) $h - 5$
    (C) $h + 2$
    (D) $h + 5$
    (E) $h + 7$

14. Which expression represents twice the product of y and 7?
    (A) $2y + 7$
    (B) $7y \div 2$
    (C) $2 \div 7y$
    (D) $2 - 7y$
    (E) $2 \times 7y$

15. Which expression represents the product of p and 6 less than p?
    (A) $p + p - 6$
    (B) $p - p - 6$
    (C) $6(p + p)$
    (D) $p(p - 6)$
    (E) $p(6 - p)$

16. Which expression is equivalent to $(8g \times 3) + (8g \times 7)$?
    (A) $8g^2 \times 3 \times 7$
    (B) $8g(3 + 7)$
    (C) $8g \times 21$
    (D) $8g + 10$
    (E) $8g(3 \times 7)$

17. Which expression is equivalent to $18 - z + z + z + z$?
    (A) $18 - 4z$
    (B) $18 + 2z$
    (C) $18 - z^4$
    (D) $18 - z^2$
    (E) $22z$

18. Which expression is equivalent to $70a - 14$?
    (A) $0.1(7a - 1.4)$
    (B) $10(700a - 14)$
    (C) $0.1(70a - 14)$
    (D) $0.1(700a - 140)$
    (E) $7a - 1$

19. Which expression represents the product of 3 and the difference of 8 and a number, n?
    (A) $24 - n$
    (B) $3n - 8$
    (C) $3(8 - n)$
    (D) $8n - 3$
    (E) $n(8 - 3)$

20. Which expression is equivalent to 47 − (p − 47+p)?
    (A) 94 + 2p
    (B) 94 − 2p
    (C) 94
    (D) 2p
    (E) p

21. Which expression represents the three times the sum of a number k and 4?
    (A) 3(4)+k
    (B) 3(k) +4
    (C) 3(k+4)
    (D) 3+(4k)
    (E) 3k − 4

22. Which is expression is equivalent to (5d × 9) + (5d × 7)?
    (A) $5d^2 \times 9 \times 7$
    (B) 5d(9 + 7)
    (C) 5d × 63
    (D) 5d + 16
    (E) 5d(9 ÷ 7)

Solving Equations and Inequalities

1. 26 − y = 14
   What is the value of y?
   (A) −14
   (B) −12
   (C) 12
   (D) 14
   (E) 40

2. 12 + 4r = −4
   What is the value of r?
   (A) −8
   (B) −4
   (C) 2
   (D) 4
   (E) 8

3. If $*p = \frac{p+2}{p-2}$, find the value of *p when p = 6.
   (A) $\frac{1}{2}$
   (B) 1
   (C) $\frac{3}{2}$
   (D) 2
   (E) 3

4. Solve for v: −6 − 2v = 12
   (A) −18
   (B) −9
   (C) 6
   (D) 9
   (E) 18

5. Which equation represents the following: 2 less than the quotient of x and 10?
   (A) 2 − 10x
   (B) 10x − 2
   (C) $\frac{x-2}{10}$
   (D) $2 - \frac{x}{10}$
   (E) $\frac{x}{10} - 2$

6. Shan bakes 30 cookies and gives x cookies to friends and relatives. Which expression represents the number of cookies Shan has left?
   (A) $\frac{x}{30}$
   (B) x − 30
   (C) 30x
   (D) 30 − x
   (E) 30 + x

7. Solve for g, if g > 0: $g^2 + 3 = 28$
   (A) −5
   (B) 5
   (C) 6
   (D) $\frac{25}{2}$
   (E) 25

8. Which of the following is equivalent to $m(3m + 3)$?
   (A) $4m + 3$
   (B) m
   (C) $m^2 + m$
   (D) $6m^2$
   (E) $3m^2 + 3m$

9. If $j + 80 = 220$, $j - 60 =$
   (A) 80
   (B) 100
   (C) 140
   (D) 160
   (E) 240

10. $30 < n^2 < 100$
    Which of the following can NOT be a value of n?
    (A) 6
    (B) 7
    (C) 8
    (D) 9
    (E) 10

11. If $j = 4$ and $k = -3$, then $jk - (j + k) =$
    (A) -19
    (B) -13
    (C) -11
    (D) 5
    (E) 19

12. Luanne has 25 candies. She then buys x more candies and gives 7 to a friend. How many candies, in terms of x, does Luanne now have?
    (A) 18
    (B) $18 - x$
    (C) $18 + x$
    (D) $25 + x$
    (E) $32 + x$

13. Which of the following is equivalent to $p(2p + 6 - 3p)$?
    (A) $2p^2 + 6 - 3p$
    (B) $6 - p$
    (C) $5p$
    (D) $6p - p$
    (E) $6p - p^2$

14. If $\frac{(g-h)}{3} = 8$ and $h = 12$, then $g =$
    (A) -12
    (B) -4
    (C) 12
    (D) 23
    (E) 36

15. If $\frac{40}{x} = 5$, what is the value of $\frac{x+4}{2}$?
    (A) 4
    (B) 6
    (C) 8
    (D) 10
    (E) 12

16. $81 - x = 13$
    What is the value of y?
    (A) -94
    (B) -68
    (C) 68
    (D) 72
    (E) 94

17. $24 + 3k = -12$
    What is the value of r?
    (A) -12
    (B) -4
    (C) 4
    (D) 6
    (E) 12

18. If $*h = \frac{h+7}{h-3}$, find the value of $*h$ when $h = 1$.
    (A) -4
    (B) $-\frac{7}{3}$
    (C) $-\frac{1}{4}$
    (D) $\frac{1}{4}$
    (E) 4

19. Which equation represents the following: 3 less than the product of x and 5?
    (A) $3 - 5x$
    (B) $5x - 3$
    (C) $\frac{x-3}{5}$
    (D) $3 - \frac{x}{5}$
    (E) $\frac{x}{15} - 3$

20. Georgina makes 27 bracelets and gives *b* bracelets to friends and relatives. Which expression represents the number of bracelets Shan has left?
    (A) $\frac{b}{27}$
    (B) $b - 27$
    (C) $27b$
    (D) $27 - b$
    (E) $27 + b$

21. Solve for *m*, if $m > 0$: $m^2 + 16 = 25$
    (A) −3
    (B) −1
    (C) 1
    (D) 3
    (E) 9

22. Which of the following is equivalent to $g(2g - 3)$?
    (A) $4g - 3$
    (B) g
    (C) $6g^2$
    (D) $2g^2 - g$
    (E) $2g^2 - 3g$

23. Nessie has 14 flowers. She then buys *f* more flowers and gives 6 to a friend. How many flowers, in terms of *f*, does Luanne now have?
    (A) 8
    (B) $8 - f$
    (C) $8 + f$
    (D) $14 - f$
    (E) $14 + f$

Multi-Step Word Problems

1. A fruit stand carries 3 times as many apples as oranges and half as many oranges as bananas. In total, the stand has 72 apples, oranges, and bananas in stock. How many of the fruits are bananas?
    (A) 6
    (B) 9
    (C) 12
    (D) 24
    (E) 36

2. Lin originally has $300. Every day, he earns $50 and spends $25. After 6 days, he gives one third of all the money he has to his sister. How much money does he now have?
    (A) $150
    (B) $250
    (C) $300
    (D) $325
    (E) $400

3. A table is *n* feet long and *n* + 8 feet wide. If the width of the table is 10 feet, what is the area, in square feet?
    (A) 16
    (B) 20
    (C) 80
    (D) 144
    (E) 180

4. Abby, Bess, and Carlo all bought muffins at a bakery. Abby bought two more than Carlo. Bess bought six fewer than Abby. If Carlo bought 7 muffins, how many muffins did Bess buy?
    (A) 1
    (B) 3
    (C) 5
    (D) 6
    (E) 15

5. Jessica has *s* slices of pie. Each slice has 3*s* cherries. How many slices of pie does she have if each slice has 18 cherries?
    (A) 3
    (B) 6
    (C) 15
    (D) 18
    (E) 54

6. In a bicycle race, Alexa rides at a constant speed of 16 miles per hour while Frank rides at 13 miles per hour. If they begin the race at the same location and time, how many miles further will Alexa have cycled than Frank after 30 minutes?
   (A) 1.5
   (B) 3
   (C) 6.5
   (D) 8
   (E) 90

7. A fifth grade class had a party, and ordered 6 pizzas, each divided into 8 slices. The fifth-grade students ate $\frac{3}{4}$ of the pizza and gave all their leftovers to the fourth-grade class. Max ate $\frac{1}{6}$ of the fourth-grade's share of the pizza. How many slices did Max eat?
   (A) 2
   (B) 3
   (C) 4
   (D) 6
   (E) 7

8. Rochelle walks her dog 1.4 kilometers to the post office, then returns home along the same route. It takes her 10 minutes to reach the post office and twice as long to return home. What is her average speed, in kilometers per hour?
   (A) 0.1
   (B) 1.4
   (C) 2.8
   (D) 4.2
   (E) 5.6

9. Dave shoveled his neighbor's driveway. He charged d dollars per hour for 3 hours of work plus a $9 equipment fee. If he made a total of $33, how much did he charge per hour?
   (A) $8
   (B) $9
   (C) $10
   (D) $11
   (E) $24

10. Sam bought four cans of soup for k dollars each. She paid with a 10-dollar bill and received $5.20 in change. How much did each can of soup cost?
    (A) $1.20
    (B) $1.30
    (C) $1.45
    (D) $4.80
    (E) $5.80

11. Trisha has s dollars in her bank account. Abe and John have a combined $7s$ dollars in their bank accounts. If Abe and John have $420 combined, how much does Trisha have in her bank account?
    (A) $6
    (B) $7
    (C) $60
    (D) $70
    (E) $80

12. Ana and Chris spent $30 together at the movies. Chris spent $4 less than what Ana spent. How much did Chris spend?
    (A) $11
    (B) $13
    (C) $17
    (D) $19
    (E) $26

13. The Spartans soccer team has m team members. The Wolves have 2 more team members than the Spartans. If they have 28 team members combined, how many team members does the Spartans have?
    (A) 12
    (B) 13
    (C) 14
    (D) 15
    (E) 26

14. Grocery store X has b bananas. Grocery store Y has $b + 6$ bananas. If Grocery store Y has 36 bananas, how many do the two stores have combined?
    (A) 30
    (B) 36
    (C) 42
    (D) 60
    (E) 66

15. Amy has *p* pencils. Liam has 6*p* pencils. If Amy has 12 pencils, how many more pencils does Liam have?
    (A) 12
    (B) 58
    (C) 60
    (D) 72
    (E) 84

16. A salad costs *d* dollars, and a sandwich costs d^2 dollars. If a sandwich costs $16, how much does a salad cost?
    (A) $2
    (B) $3
    (C) $4
    (D) $5
    (E) $8

17. A helmet costs d^2 dollars, and a bike light costs *d* dollars. If a helmet costs $36, how much more expensive is a helmet than a bike light?
    (A) $6
    (B) $18
    (C) $30
    (D) $36
    (E) $42

18. Tanya bought six iced teas for *d* dollars each. If she paid with a twenty-dollar bill and received $6.80 in change, how much did each ice tea cost?
    (A) $1.20
    (B) $1.80
    (C) $2.20
    (D) $2.30
    (E) $2.36

19. A rectangular window is *t* feet wide and *t* + 3 feet tall. If the window is 6 feet tall, what is the perimeter of the window, in feet?
    (A) 3
    (B) 6
    (C) 9
    (D) 18
    (E) 20

20. Tom has four times as many parakeets as Robert. If they have a total of 60 parakeets combined, how many does Tom have?
    (A) 12
    (B) 15
    (C) 40
    (D) 45
    (E) 48

21. Lia and Sean baked brownies to sell at a bake sale. Lia baked one third the number of brownies that Sean baked. If they baked 48 brownies combined, how many brownies did Lia bake?
    (A) 12
    (B) 16
    (C) 24
    (D) 36
    (E) 48

22. Highway 101 is *m* miles long. Highway 202 is 3*m* – 6 miles long. Highway 303 is 6*m* – 18 miles long. If Highway 202 is 6 miles long, how long are highways 101 and 303 combined?
    (A) 4 miles
    (B) 6 miles
    (C) 8 miles
    (D) 10 miles
    (E) 12 miles

23. Donny set up a lemonade stand and charged *d* dollars a cup. At the end of the day, he paid his mother back the $10 she spent on his lemonade-making supplies. If he had $50 left and sold 20 cups of lemonade, how much did he charge per cup of lemonade?
    (A) $2.00
    (B) $2.50
    (C) $3.00
    (D) $3.50
    (E) $4.00

Geometry

Pythagorean Theorem

1. Which is the missing length of the side in the right triangle shown below?

   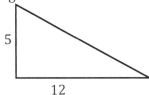

   (A) 11
   (B) 13
   (C) 15
   (D) 17
   (E) 26

2. What is the area, in square units, of the triangle shown below?

   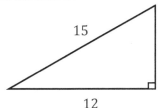

   (A) 42
   (B) 48
   (C) 54
   (D) 65
   (E) 108

3. What is the length of segment *XZ*?

   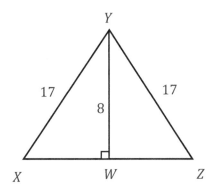

   (A) 15
   (B) 48
   (C) 30
   (D) 52
   (E) 60

4. A rectangular park measures 6 kilometers long and 8 kilometers wide. A straight path runs through the park from one corner to the opposite corner, as shown below.

   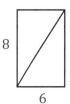

   What is the length, in kilometers, of the path?
   (A) 10
   (B) 14
   (C) 15
   (D) 17
   (E) 24

5. Two students leave the same school, with one heading west and one heading south. After one hour, the westbound student has traveled 16 km. If the two students are 20km apart along a straight line, how far, in km, has the southbound student traveled?

   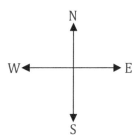

   (A) 10
   (B) 12
   (C) 18
   (D) 20
   (E) 30

6. Jonathan draws an X on a piece of rectangular paper by drawing diagonals from one corner to the opposite corner. The length of each diagonal is 25 inches. If the paper if 24 inches long, what is the width, in inches, of the box?

 24 inches

(A) 1
(B) 5
(C) 7
(D) 8
(E) 49

7. A 20-foot long ladder is leaning against a building. If the ladder reaches 16 feet up the building, how far away, in feet, is the base of the ladder from the building?

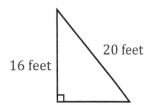

(A) 4
(B) 12
(C) 14
(D) 15
(E) 16

8. What is the length of segment AC in the figure shown below?

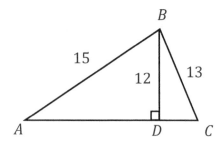

(A) 5
(B) 9
(C) 13
(D) 14
(E) 16

9. The size of television screens is determined by measuring diagonally from one corner to the opposite corner. The screen size of a certain television is determined to be 17 inches. If the width of the television is 15 inches, what is the length, in inches?

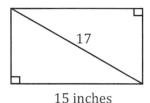

15 inches

(A) 2
(B) 5
(C) 7
(D) 8
(E) 12

Coordinate Plane

1. Which ordered pair is in the third quadrant of the coordinate plane?
   (A) (−4, 2)
   (B) (−8, −1)
   (C) (4, −3)
   (D) (1, 2)
   (E) (0, 3)

2. Which ordered pair is in the fourth quadrant of the coordinate plane?
   (A) (−3, 2)
   (B) (−1, −1)
   (C) (8, −3)
   (D) (−1, 2)
   (E) (5, 0)

3. What is the location of point R on the coordinate plane?

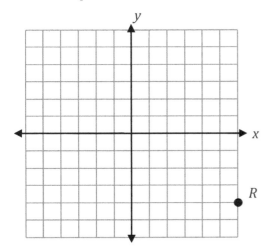

   (A) (6, 4)
   (B) (4, 6)
   (C) (−4, 6)
   (D) (−6, 4)
   (E) (6, −4)

4. Which point on the coordinate plane is located exactly 5 units from (−2, −1)?

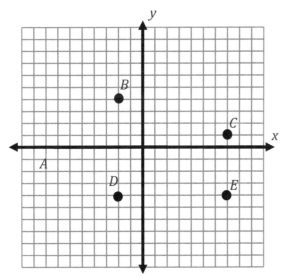

   (A) A
   (B) B
   (C) C
   (D) D
   (E) E

5. Which set of 4 points is listed in this order: Quadrant IV, Quadrant III, Quadrant II, Quadrant I?
   (A) {(2, 5), (−3, 1), (−2, −5), (3, −1)}
   (B) {(2, 5), (−3, −1), (−2, 5), (3, −1)}
   (C) {(2, −5), (−3, −1), (−2, 5), (3, 1)}
   (D) {(2, −5), (3, 1), (−2, 5), (−3, −1)}
   (E) {(2, −5), (−3, 1), (−2, −5), (3, 1)}

6. Maia made a map of her friends' houses on this coordinate plane.

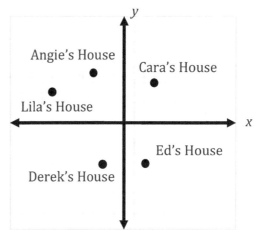

   Whose house is located at (−3, 5)?
   (A) Angie's house
   (B) Lila's house
   (C) Cara's house
   (D) Derek's house
   (E) Ed's house

7. The coordinate grid represents a town and the grid lines represent the roads. Tim wants to travel along the roads from his school to his home. Which set of directions will lead Tim from his school to his house?

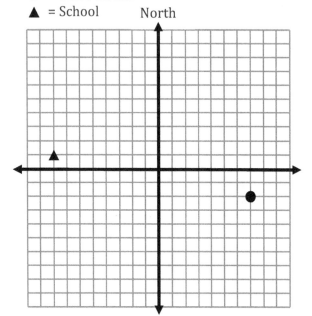

(A) Travel north for 3 units and then west for 15 units.
(B) Travel south for 1 unit and then east for 7 units
(C) Travel south for 3 units and then east for 8 units
(D) Travel south for 1 unit and then east for 8 units
(E) Travel south for 3 units and then east for 15 units

8. Point *M* is in the second quadrant and *Q* is in the third quadrant. Point *M* is exactly 6 units from point *Q* along a grid line. Which could be the coordinates of points *M* and *Q*?
(A) (3, 4) and (−3, 4)
(B) (4, 3) and (4, −3)
(C) (−4, 3) and (−4, −3)
(D) (−4, 3) and (−3, −4)
(E) (−4, 6) and (−4, −6)

9. Point *N* is located at (−9, −3). If the point is moved 5 units to the right and 6 units up, what is the new location of point *N*?
(A) (4, 3)
(B) (−4, 6)
(C) (−3, 2)
(D) (−4, 3)
(E) (4, −3)

10. Rita drew the following points on a straight line on a coordinate grid: (−6, 0), (−6, −4), (−6, 2), and (−6, 5). Which two points on Rita's line are *farthest* from one another?
(A) (−6, −4) and (−6, 5)
(B) (−6, 0) and (−6, 2)
(C) (−6, −4) and (−6, 0)
(D) (−6, 2) and (−6, 5)
(E) (−6, 0) and (−6, 5)

11. Point *G* on a coordinate plane is located at (1, 6). Point *F* is located 3 units to the left and 2 units down from point *G*. What is the location of point *F*?
(A) (4, 8)
(B) (−1, 3)
(C) (−2, 4)
(D) (−2, 8)
(E) (−2, −4)

Transformations

1. If Triangle NOP is reflected through the origin, what is the new location of vertex O?

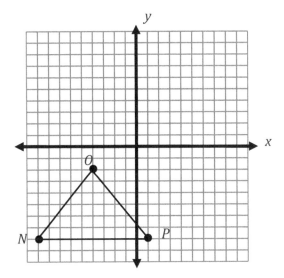

   (A) (4, 2)
   (B) (4, –2)
   (C) (–2, 4)
   (D) (2, –4)
   (E) (–4, 2)

2. If Rectangle GHIJ is reflected over the line segment IJ, what is the new location of vertex G?

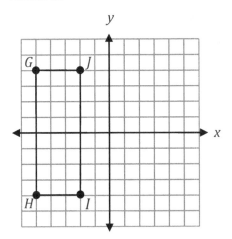

   (A) (5, 4)
   (B) (–5, –4)
   (C) (1, 4)
   (D) (1, –4)
   (E) (–2, 4)

3. Which transformation moves Line Segment VW from Position 1 to Position 2?

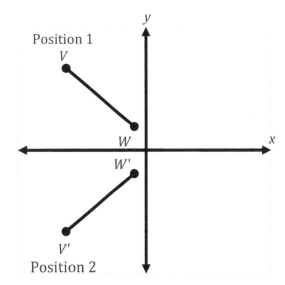

   (A) reflection across the x-axis
   (B) 90-degree clockwise rotation about Point W
   (C) 90-degree counterclockwise rotation about Point W
   (D) 180-degree rotation about Point V
   (E) translation 4 units down

4. Which translation moves a point from (4, –3) to (–2, 5)?
   (A) 2 units left, 2 units up
   (B) 2 units left, 2 units down
   (C) 6 units left, 8 units up
   (D) 6 units right, 8 units down
   (E) 8 units right, 6 units down

5. Vertex J of Rectangle JKLM is located at (–6, 3), vertex K is located at (–6, –2), vertex L is located at (3, –2) and vertex M is located at (3, 3). If the rectangle is translated down 4 units, which quadrant(s) will NOT contain the rectangle?
   (A) I and II
   (B) I and IV
   (C) I only
   (D) III and IV
   (E) II only

6. If parallelogram QRST is rotated 90 degrees clockwise about Point R, what are the new coordinates of S?

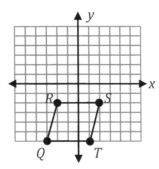

(A) (−1, −6)
(B) (−2, −2)
(C) (2, −2)
(D) (2, −6)
(E) (−2, −6)

7. A triangle, PQR, is shown in Position 1 and Position 2. What series of transformations could move the triangle from Position 1 to Position 2?

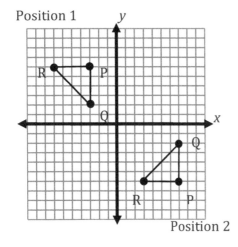

(A) reflection across x-axis, reflection across y-axis
(B) reflection across y-axis, reflection across x-axis
(C) 90-degree rotation counterclockwise, reflection across x-axis, translation 10 units right
(D) 90-degree rotation clockwise, reflection across y-axis, translation 6 units down
(E) reflection across x-axis, translation 10 units right

8. Triangle XYZ is reflected across the y-axis, and rotated 90 degrees clockwise about Point Z. Which of the following shows the new position of Triangle XYZ?

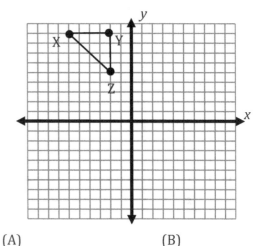

(A)   (B)

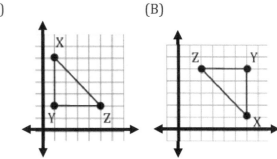

(C)   (D)

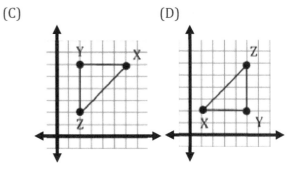

(E)

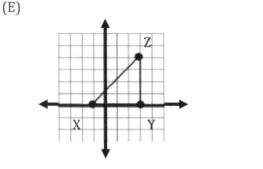

9. Triangle *JKL* is rotated 90 degrees clockwise about point *K* and reflected across the new line segment formed by *KL*. Which figure shows the triangle's new location?

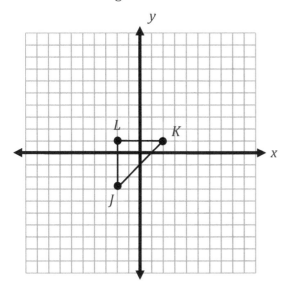

(A)

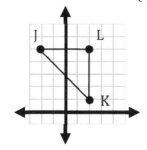

(B)

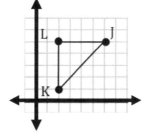

(C)

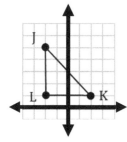

(D)

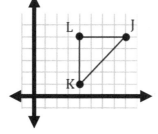

(E)

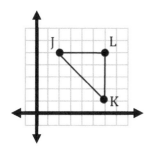

10. What does the shape shown below look like after it is rotated 90 degrees counterclockwise?

(A)

(B)

(C)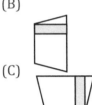

(D)

(E)

11. Gia designed the shape shown below for a collage. Which of the following is NOT a possible rotation for Gia's shape?

(A)

(B)

(C)

(D)

(E)

Circles

1. If the circumference of a certain circle is 10π, what is the area of the circle?
   (Note: A = πr² and C = πd)
   (A) 100π
   (B) 20π
   (C) 25π
   (D) 50π
   (E) 5π

2. The area of a square piece of paper is 64 square inches. If a circle is to be cut from the paper, what is the largest possible area the circle can have?
   (A) 8π in²
   (B) 16π in²
   (C) 24π in²
   (D) 32π in²
   (E) 64π in²

3. What is the circumference, in centimeters, of the largest circle that can fit inside a square whose perimeter is 48 centimeters?
   (Note: P = 4s and C = πd)
   (A) 12
   (B) 36
   (C) 12π
   (D) 36π
   (E) 144π

4. Simon's art class was given the task of cutting out circles from square pieces of paper. If the area of Simon's piece of paper is 36 square inches, which of the following cannot be the circumference of a circle cut from the paper?
   (A) 6 inches
   (B) 3π inches
   (C) 4π inches
   (D) 6π inches
   (E) 7π inches

5. A bin at a toy shop displays twelve beach balls. There are two balls across the front, three balls across the side and two balls high. If each ball has a radius of 9 inches, what are the dimensions, in feet, of the smallest bin that can contain the balls?
   (A) 1.5 × 2.25 × 1.5
   (B) 1.8 × 2.7 × 1.8
   (C) 2 × 3 × 2
   (D) 3 × 4.5 × 3
   (E) 3.6 × 5.4 × 3.6

6. If the radius of the circle below is 5 cm, what is the area of the shaded region, in terms of π? *(Note: A = πr²)*

   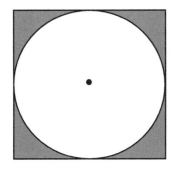

   (A) 100π − 25 cm²
   (B) 100π − 100 cm²
   (C) 100 − 5π cm²
   (D) 100 − 25π cm²
   (E) 25π − 9 cm²

7. Eight balls will be placed in a cube box for shipment. If the radius of each ball is 3 inches, what are the dimensions, in inches, of the smallest box that can contain them?
   (A) 2 × 2 × 2
   (B) 3 × 3 × 3
   (C) 6 × 6 × 6
   (D) 3π × 3π × 3π
   (E) 12 × 12 × 12

8. A circle is drawn such that each point on the circumference of the circle is 4 inches from the center. Which of the following *cannot* be the length of a line segment whose endpoints are on the circumference of the circle?
   (A) 3
   (B) 4
   (C) 6
   (D) 8
   (E) 12

9. The area of a circle is between 30π cm² and 40π cm². Which could be the length of the circle's diameter?
   (A) 6 cm
   (B) 12 cm
   (C) 15 cm
   (D) 20 cm
   (E) 36 cm

10. Use the following diagram to answer the question.

    Line segment *AB* has endpoints on the circle. *AB* has a length of 6 and it does not pass through the center of the circle.

    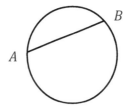

    Which of the following could be the circumference of the circle? *(Note: C = πd)*
    (A) 3π
    (B) 4π
    (C) 5π
    (D) 6π
    (E) 7π

11. Which of the following *cannot* be the area of the circle?
    (A) 9π
    (B) 12π
    (C) 16π
    (D) 25π
    (E) 36π

Two and Three Dimensional Shapes

1. A shape has four sides. Which of the following MUST be true?
   (A) The sum of the interior angles is 180°.
   (B) The shape is a rectangle.
   (C) No interior angle measures greater than 90°.
   (D) At least two sides are equal in length.
   (E) The sum of the interior angles is 360°.

2. How many vertices does a pentagonal prism have?
   (A) 5
   (B) 6
   (C) 7
   (D) 10
   (E) 14

3. How many edges does a heptagonal prism have? (A heptagon is a polygon with 7 sides).
   (A) 6
   (B) 12
   (C) 14
   (D) 18
   (E) 21

4. Which of the following has exactly five faces?
   (A) cube
   (B) pentagon
   (C) rectangular prism
   (D) triangular pyramid
   (E) triangular prism

5. Which of the following does NOT have exactly six faces?
   (A) triangular prism
   (B) rectangular prism
   (C) pentagonal pyramid
   (D) trapezoidal prism
   (E) cube

6. Which of the following could NOT describe at least one of the faces of a trapezoidal prism?
   (A) trapezoid
   (B) rectangle
   (C) parallelogram
   (D) isosceles triangle
   (E) square

7. Which figure can be constructed from this net?

   (A)
   (B)
   (C)
   (D)
   (E)

8. A polygon on the coordinate plane has vertices at these coordinates:
   (2, 3) (1, −1) (4, 3) (5, −1)
   Which of the following could be the polygon described by vertices at these coordinates?
   (A)
   (B)
   (C)
   (D)
   (E)

9. The figure below is a trapezoidal prism. A cross section is taken at the center of the prism, perpendicular to its base. What is the best description of the shape of this cross section?

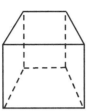

(A) trapezoid
(B) isosceles triangle
(C) pentagon
(D) right triangle
(E) parallelogram

10. A figure can be constructed from this net. How many edges will this figure have?

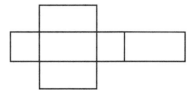

(A) 6
(B) 8
(C) 12
(D) 16
(E) 19

Spatial Reasoning

1. Which of the following shapes can be used to create the figure shown without overlap?

(A)

(B)

(C)

(D)

(E)

2. Jordan built the columns below by stacking cubes on top of and next to one another. There are no gaps between the cubes. What is the total number of cubes in Jordan's columns?

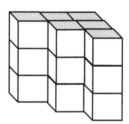

(A) 6
(B) 9
(C) 12
(D) 18
(E) 24

3. In which of the following figures will two quadrilaterals be formed if Q and R are joined by a straight line?

(A)

(B)

(C)

(D)

(E)

4. An equilateral triangle is drawn. Its vertices are A, B, and C. The midpoint of AB is D, the midpoint of AC is F, and the midpoint of BC is E. How many different triangles can be drawn with 3 vertices chosen from A, B, C, D, E, and F?

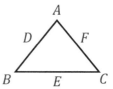

(A) 6
(B) 17
(C) 20
(D) 216
(E) 729

5. Dan lives 8 miles south and 15 miles east of his brother Dave. If there were a straight path paved from Dan's house to Dave's, how long would it be?
(A) 7 miles
(B) 12.5 miles
(C) 15 miles
(D) 17 miles
(E) 23 miles

6. How many rectangles are in the figure below? (They do not have to be 1 × 1.)

(A) 9
(B) 14
(C) 18
(D) 24
(E) 36

Measurement

Time and Money

1. It took Adela $1\frac{1}{4}$ hours to run around a lake. If she finished her run at 7:10 p.m., when did she start running?
    (A) 5:45 p.m.
    (B) 5:50 p.m.
    (C) 5:55 p.m.
    (D) 6:00 p.m.
    (E) 6:05 p.m.

2. Lana has $26. If pencils cost 30¢ each, what is the greatest number of pencils Lana can buy with her money?
    (A) 0
    (B) 1
    (C) 26
    (D) 86
    (E) 87

3. In a jar, there are ten of each of the following coins: 1¢, 5¢, 10¢, and 25¢. If Regina needs exactly $0.71, what is the least number of coins she must take from the jar?
   (A) 3
   (B) 4
   (C) 5
   (D) 6
   (E) 7

4. One apple can be traded for 4 clementines. Sixteen clementines can be traded for 26 figs. How many figs can be traded for 6 apples?
   (A) 13
   (B) 26
   (C) 39
   (D) 78
   (E) 156

5. Albert's cable plan costs $55 per month. Premium channels cost an additional $2.50 each per month. If Albert pays $72.50 per month for cable, how many premium channels does he pay for?
   (A) 4
   (B) 5
   (C) 6
   (D) 7
   (E) 29

6. The person who came in 3rd place in a triathlon finished at 3:05 p.m. The person who came in 2nd place finished 20 minutes before that. The person who came in 1st place finished $\frac{1}{4}$ of an hour before the person who came in 2nd place. At what time did the person who came in 1st place finish?
   (A) 2:15 p.m.
   (B) 2:20 p.m.
   (C) 2:25 p.m.
   (D) 2:30 p.m.
   (E) 2:35 p.m.

7. Charlie charges a flat rate of $26 and an additional fee of $6.50 for every hour he spends mowing the lawn. On Saturday, he earned $65 mowing his neighbor's lawn. How many hours did Charlie spend to do it?
   (A) 2
   (B) 4
   (C) 5
   (D) 6
   (E) 10

8. Kristina is buying bookshelves from a furniture store. She has $6,000 to spend, and bookshelves cost $350 each. What is the greatest number of bookshelves she can buy?
   (A) 15
   (B) 16
   (C) 17
   (D) 18
   (E) 19

9. A store sells hats and scarves. One day, they sold 7 hats at a price of $14.50 each. They also sold 20 scarves. If they earned a total of $401.50, how much does one scarf cost?
   (A) $7.50
   (B) $14.50
   (C) $15.00
   (D) $19.35
   (E) $22.00

10. Tickets for a concert normally cost $32.50 each. Groups of 20 or more receive a discounted price. A group of 22 paid $545.25 total for their tickets. How much was their discount?
    (A) $104.75
    (B) $159.75
    (C) $169.75
    (D) $170.75
    (E) $179.75

Area, Perimeter and Volume

1. Karl wants to surround his garden with a chicken wire fence. One side of the garden has a length of 6 m and the area of the garden is 48 m². How many meters of chicken wire fencing does Karl need?
   (A) 8
   (B) 14
   (C) 28
   (D) 48
   (E) 108

2. What is the perimeter of the shape shown below?

   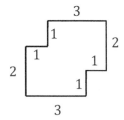

   (A) 12
   (B) 14
   (C) 16
   (D) 20
   (E) 24

3. Bianca's rectangular area rug is 5 feet long and 36 inches wide. What is its area in square feet? *(Note: 1 ft = 12 in)*
   (A) 15 ft²
   (B) 41 ft²
   (C) 82 ft²
   (D) 180 ft²
   (E) 2,160 ft²

4. A rectangle is formed from 4 congruent triangles. If each triangle has a base of 8 mm and a height of 8 mm, what is the area of the rectangle? *(Note: Area of a triangle = $\frac{1}{2}bh$)*
   (A) 16 mm²
   (B) 32 mm²
   (C) 64 mm²
   (D) 128 mm²
   (E) 256 mm²

5. If exactly half of a pan measuring 10 in × 8 in × 3 is filled with cake batter, what is the volume of the cake batter?
   (A) 60 in³
   (B) 80 in³
   (C) 120 in³
   (D) 240 in³
   (E) 320 in³

6. In the figure below, *JH* is the height of the triangle and measures 8 cm, *GI* = 5 cm, and *GH* = *HI*. What is the area of triangle *JHI*?

   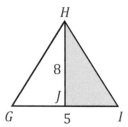

   (A) 10 cm²
   (B) 13 cm²
   (C) 20 cm²
   (D) 21 cm²
   (E) 40 cm²

7. Kelsey created a shape by placing a rectangle on top of a triangle as shown below. *VW* = *WX* = *WY* = 5, and *SY* = *YT* = 3. If there are no spaces between the rectangle and triangle and they do not overlap, what is the area of Kelsey's shape?

   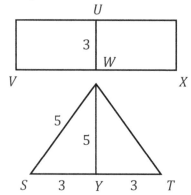

   (A) 30
   (B) 45
   (C) 60
   (D) 75
   (E) 90

Use the following figure to answer questions 8 and 9.

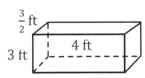

8. Yolanda is moving and packing her books in large boxes like the one above. What is the volume of the box?
   (A) $7\frac{1}{2}$ ft³
   (B) 12 ft³
   (C) 18 ft³
   (D) 36 ft³
   (E) 45 ft³

9. What is the surface area of the box? (Note: Surface area = 2lw + 2lh + 2wh)
   (A) 15 ft²
   (B) 27 ft²
   (C) 36 ft²
   (D) 45 ft²
   (E) 48 ft²

10. In the figure below, UVWX is a square, VU = 4 cm, WY = 5 cm, and XY = $\frac{3}{4}$ UX. (Note: Figure is not drawn to scale.)

    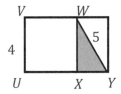

    What is the perimeter of triangle WYX?
    (A) 9 cm
    (B) $10\frac{3}{4}$ cm
    (C) 12 cm
    (D) 13 cm
    (E) 14 cm

11. Vinnie is laying tiles on his kitchen floor. The floor measures 8 ft. × 9 ft. and each square tile is 6 inches long. How many tiles does he need?
    (A) 2
    (B) 12
    (C) 144
    (D) 288
    (E) 432

12. A playground covers 400 square meters. If it contains a 3 m × 5 m sandbox and a 15 m × 8 m volleyball court, what is its perimeter?
    (A) 70 m
    (B) 80 m
    (C) 265 m
    (D) 400 m
    (E) It cannot be determined from the information provided.

13. A rectangle has an area of 72 in². If the rectangle's length and width are whole numbers, which of the following CANNOT be its perimeter?
    (A) 34 in
    (B) 36 in
    (C) 76 in
    (D) 144 in
    (E) 146 in

14. Every angle in the figure below is a right angle. What is the perimeter of the figure?

    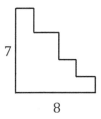

    (A) 15
    (B) 30
    (C) 56
    (D) 112
    (E) It cannot be determined from the information provided.

15. Yasmin has a 4 in × 4 in × 4 in cube made of 4 smaller cubes along each side as shown below. If she breaks the large cube into 4 equally sized pieces, what is the volume of each piece?

    (A) 1 in³
    (B) 4 in³
    (C) 8 in³
    (D) 16 in³
    (E) 64 in³

16. The below figure is a rectangular prism. If N is the midpoint of MO, MN = 5, NPQO is a square, and QR = 6, what is the area of triangle PQR?

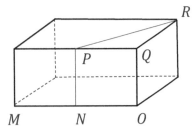

(A) 12.5
(B) 15
(C) 25
(D) 30
(E) 60

17. Derek's rectangular swimming pool is one-third filled with water. If the dimensions of his pool are 15 feet wide, 25 feet long, and 4 feet deep, what is the volume of the water in the pool, in cubic feet?
(A) $\frac{1,500}{27}$
(B) 500
(C) 750
(D) 1,500
(E) 4,500

Angles

1. Angles G and H are supplementary. Angle G measures 54°. What is the measure of angle H?
   (A) 36°
   (B) 92°
   (C) 126°
   (D) 136°
   (E) 306°

2. The sum of angle J and angle K is 145°. If angle J measures 64°, what is the measure of angle K?
   (A) 35°
   (B) 45°
   (C) 72°
   (D) 81°
   (E) 116°

3. In the triangle shown, what is the value of x?

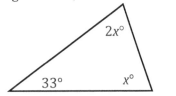

   (A) 33°
   (B) 45°
   (C) 49°
   (D) 98°
   (E) 147°

4. In the figure below, two lines intersect as shown.

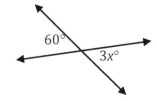

   What is the value of x?
   (A) 20
   (B) 30
   (C) 60
   (D) 72
   (E) 80

5. In the triangle shown, what is the measure of angle B?

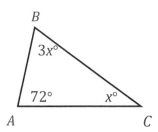

   (A) 27°
   (B) 36°
   (C) 45°
   (D) 81°
   (E) 108°

6. The figure below is composed of two isosceles triangles. What is the measure of angle CDA?

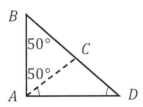

(A) 120°
(B) 100°
(C) 80°
(D) 50°
(E) 40°

7. The shape below is a rhombus.

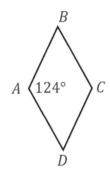

What is the measure of angle ABC?
(A) 28°
(B) 56°
(C) 64°
(D) 112°
(E) 248°

8. The measure of angle BAD is 130°. The measure of angle CAD is 98°.

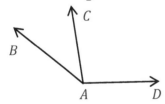

What is the measure of angle BAC?
(A) 90°
(B) 82°
(C) 64°
(D) 32°
(E) 22°

9. The figure below is composed of two isosceles triangles. What is the measure of angle CDA?

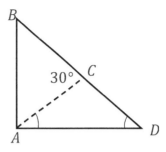

(A) 15°
(B) 30°
(C) 75°
(D) 80°
(E) 150°

10. The shape below is a rhombus.

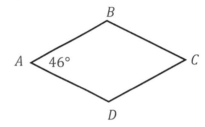

What is the measure of angle ADC?
(A) 44°
(B) 94°
(C) 104°
(D) 134°
(E) 268°

11. In the figure shown, line c intersects parallel lines a and b. What is the value of x?

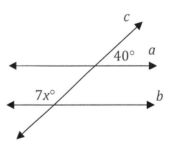

(A) 10
(B) 14
(C) 20
(D) 70
(E) 140

12. In the figure shown, line *a* is parallel to line *b*, and lines *c* and *d* intersect lines *a* and *b* as shown. What is the value of *x*?

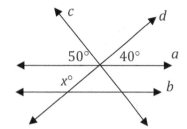

(A) 40
(B) 50
(C) 60
(D) 130
(E) 140

13. In the figure shown, line *a* is parallel to line *b* and lines *c* and *d* intersect lines *a* and *b*. What is the value of *x*?

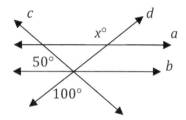

(A) 30
(B) 50
(C) 100
(D) 135
(E) 150

Unit Analysis

1. Chris's science book has a mass of 2.1 kilograms. His math book has a mass of 3.2 kilograms. What is the difference, in grams, between the masses of the two books?
   (A) 11
   (B) 100
   (C) 1,100
   (D) 1,200
   (E) 2,000

2. Tom ran a mile in $7\frac{1}{2}$ minutes. Allie ran the same distance. Her time was 45 seconds longer than Tom's. How long did it take Allie to run a mile?
   (A) 7 minutes and 45 seconds
   (B) 8 minutes
   (C) 8 minutes and 15 seconds
   (D) 8 minutes and 30 seconds
   (E) 8 minutes and 45 seconds

3. Samantha bought four 350-milliliter bottles of apple juice. How many liters of juice did Samantha buy altogether?
   (A) 0.14
   (B) 0.35
   (C) 1.4
   (D) 3.5
   (E) 140

4. How many pints is 2.5 gallons, if 1 gallon is equivalent to 8 pints?
   (A) 10
   (B) 18
   (C) 20
   (D) 22
   (E) 24

5. Theresa's backpack weighed 1,384 grams. What does the backpack weigh in pounds? *(Note: 1 pound = 346 grams).*
   (A) 4
   (B) 40
   (C) 86
   (D) 400
   (E) 478

6. There are 8 pencils in a pencil holder. Each pencil has a mass of 40 grams. The total mass of the pencils and the holder together is 0.6 kilograms. What is the mass of the pencil holder, in grams?
   (A) 28
   (B) 280
   (C) 300
   (D) 560
   (E) 5,680

7. At a long jump competition, Maya jumped 20 feet and 4 inches, while John jumped 22 feet, 6 inches. How much farther, in inches, did John jump than Maya?
   (Note: 1 foot = 12 inches)
   (A) 2
   (B) 10
   (C) 20
   (D) 24
   (E) 26

8. A recipe calls for 2 cups of sugar. Julio has a 4-pound bag of sugar. If 2 cups of sugar weigh 1 pound, what fraction of the bag of sugar does Julio need for the recipe?
   (A) $\frac{1}{8}$
   (B) $\frac{1}{4}$
   (C) $\frac{1}{2}$
   (D) $\frac{3}{8}$
   (E) $\frac{3}{4}$

9. The length of Emily's desk is 76.2 centimeters. What is the length of Emily's desk in inches?
   (Note: 1 inch = 2.54 centimeters)
   (A) 28
   (B) 30
   (C) 67
   (D) 128
   (E) 194

10. Mr. Lopez bought a sofa that is 72 inches long. What is this length in meters?
    (Note: 1 inch = 2.54 centimeters)
    (A) 1.8288
    (B) 2.834
    (C) 18.288
    (D) 28.456
    (E) 182.88

Data Analysis

Interpreting Bar Graphs

1. The graph below shows the number of plants in a garden. How many total spinach and tomato plants are there in the garden?

   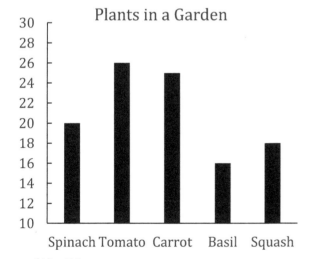

   (A) 20
   (B) 26
   (C) 36
   (D) 46
   (E) 50

2. The graph below shows the result of a survey to determine favorite sports. Which sports combined received the same number of votes as baseball?

   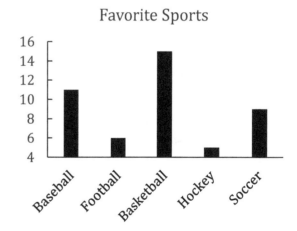

   (A) football and basketball
   (B) football and hockey
   (C) football and soccer
   (D) basketball and hockey
   (E) hockey and soccer

3. The graph below shows the sales of school supplies at an office supply store. Which is the best estimate for how many more notebooks were sold than folders?

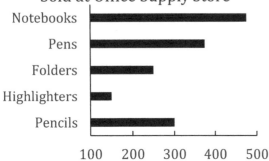

(A) 150
(B) 225
(C) 300
(D) 450
(E) 475

4. The bar graph below shows the number of children who participated in activities at a summer camp. How many more girls participated in swimming than boys participated in canoeing?

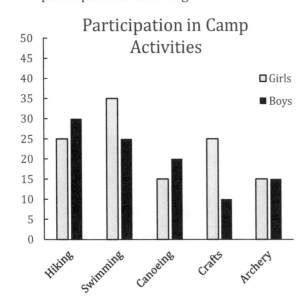

(A) 5
(B) 10
(C) 15
(D) 20
(E) 35

5. The graph below shows the results of a survey of what kind of pets people have. How many more people have dogs than have fish and other pets combined?

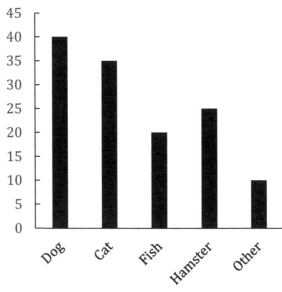

(A) 10
(B) 15
(C) 20
(D) 25
(E) 30

Use the following information and graph to answer questions 6 and 7.

The bar graph below shows the results of a survey for how many students in a school speak different languages.

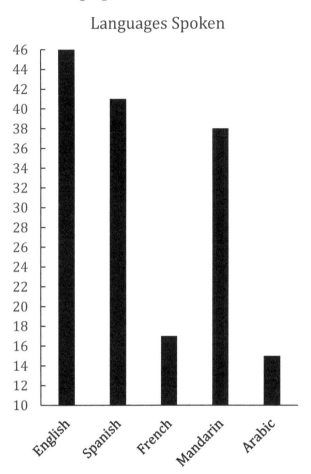

6. How many more students speak Spanish than Arabic?
   (A) 16
   (B) 18
   (C) 25
   (D) 26
   (E) 30

7. How many more students speak English than speak French and Arabic combined?
   (A) 14
   (B) 15
   (C) 17
   (D) 32
   (E) 40

Use the following information and graph to answer questions 8 and 9.

The bar graph below shows a city's high and low temperatures, in degrees Fahrenheit, over a five-day period.

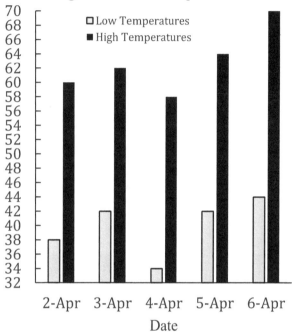

8. On which day was the difference between the high and low temperatures the same as the difference recorded on April 2?
   (A) April 3
   (B) April 4
   (C) April 5
   (D) April 6
   (E) It cannot be determined from the information given.

9. Which day had the greatest difference in its high and low temperatures?
   (A) April 2
   (B) April 3
   (C) April 4
   (D) April 5
   (E) April 6

Use the following information and graph to answer questions 10 and 11.

The bar graph below shows the number of devices, by type, sold at an electronics store last year.

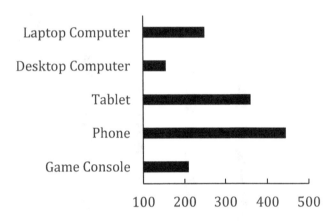

Number and Type of Devices Sold at Electronics Store

10. Approximately how many more phones than laptop computers were sold?
    (A) 100
    (B) 150
    (C) 200
    (D) 275
    (E) 300

11. Approximately how many laptop and desktop computers were sold altogether?
    (A) 150
    (B) 250
    (C) 300
    (D) 400
    (E) 450

Interpreting Histograms

Use the following graph to answer questions 1 and 2.

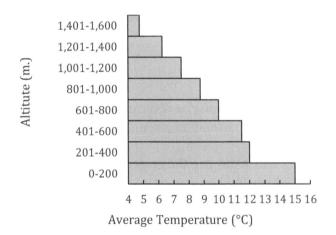

Temperature and Altitude for Mt. Kina

1. On a chilly day, a team of meteorologists records the temperature at different altitudes on Mt. Kina. What is the difference between the average temperature at 0-200 meters and 801-1000 meters?
   (A) 4.75°C
   (B) 6.25°C
   (C) 7.25°C
   (D) 8.75°C
   (E) 22.5°C

2. Which statement is true based on the data in the graph?
   (A) As altitude increases, temperature increases.
   (B) As temperature increases, altitude increases.
   (C) As altitude increases, temperature decreases.
   (D) As temperature decreases, altitude decreases.
   (E) There is no relationship between temperature and altitude

Use the following information and graph to answer questions 3 and 4.

Shira collected insects, then measured and identified them for her science class. She recorded the data in the histogram below.

Shira's Insect Collection

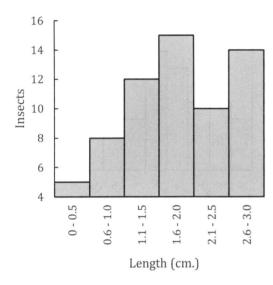

3. What fraction of Shira's insects were between 0.6 and 1 cm in length?
   (A) $\frac{1}{8}$
   (B) $\frac{1}{4}$
   (C) $\frac{3}{8}$
   (D) $\frac{5}{8}$
   (E) 8

4. How many insects were exactly 1.6 cm long?
   (A) 0
   (B) 8
   (C) 12
   (D) 15
   (E) It cannot be determined from the information provided.

Use the following information and graph information to answer questions 5 and 6.

Donut Zone is a regional chain that sells breakfast foods and coffee. They recorded their sales in the histogram below.

2017 Donut Sales

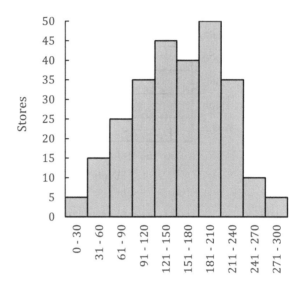

5. How many stores sell $121-$210 million in products?
   (A) 35
   (B) 45
   (C) 80
   (D) 90
   (E) 135

6. Each Donut Zone store needs $120 million to make a profit. How many stores made no more than half that amount?
   (A) 13
   (B) 20
   (C) 35
   (D) 80
   (E) It cannot be determined from the information provided.

Use the following graph to answer questions 7 and 8.

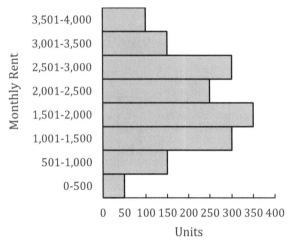

7. What is the ratio of units for $0-1,000 to units for $1,501-$2,500?
   (A) 1:7
   (B) 1:6
   (C) 1:3
   (D) 4:7
   (E) 4:5

8. If 75 units with a rental price of $1,750 and 90 units with a rental price of $2,960 are taken off the market, what price range has the largest number of available units?
   (A) $501-$1,000
   (B) $1,001-$1,500
   (C) $1,501-$2,000
   (D) $2,001-$2,500
   (E) $2,501-$3,000

Use the following information and graph to answer questions 9 and 10.

Pedro asks a randomly chosen group of his 6th grade classmates how many times they checked their phone the previous day. He recorded his data in the histogram below.

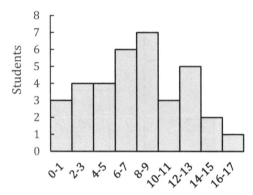

9. How many students checked their phone more than five times?
   (A) 10
   (B) 17
   (C) 24
   (D) 28
   (E) 35

10. How many students participated in the survey?
    (A) 8
    (B) 9
    (C) 17
    (D) 35
    (E) 36

Interpreting Line Graphs

1. Bartonville is a "boom town" with a rapidly increasing population. The graph below shows the population growth over several years.

   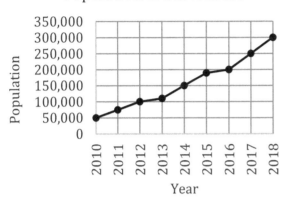

   How many fewer people lived in Bartonville in 2012 than in 2018?
   (A) 50,000
   (B) 100,000
   (C) 200,000
   (D) 300,000
   (E) 400,000

2. Lee's Hardware sells fans during the summer. Every day at 6:30 PM, after the store closes, the department manager counts the fans on the shelf. He records the data on the graph below. If the store was only restocked before opening on Friday, how many fans did it sell between Monday and Saturday?

   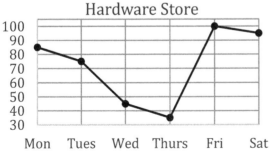

   (A) 5
   (B) 15
   (C) 50
   (D) 55
   (E) 65

3. Harriet recently took a scenic road trip. The graph below shows how far she traveled each hour. What is her average speed between 12:00 and 5:00 PM?

   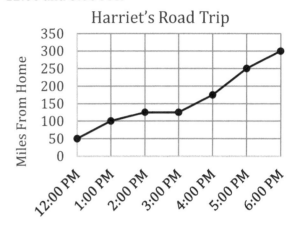

   (A) 40 mph
   (B) 45 mph
   (C) 50 mph
   (D) 55 mph
   (E) 60 mph

Use the following graph to answer questions 4 and 5.

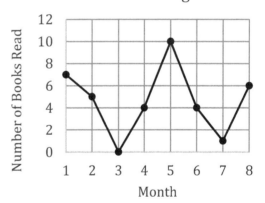

4. Anna's class is awarding prizes for reading books. Anna counts how many books she reads every month for 8 months. How many books did she read in total over the 8 months?
   (A) 6
   (B) 31
   (C) 36
   (D) 37
   (E) 38

5. Anna's classmate Jenny read exactly 4 books every week. During how many weeks did Anna read more books than Jenny?
   (A) 3
   (B) 4
   (C) 5
   (D) 6
   (E) 7

Use the following graph to answer questions 6 and 7.

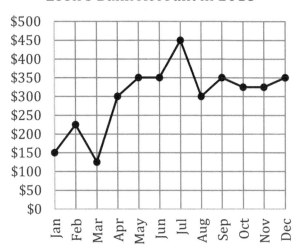

6. Leon monitors his bank balance on the first day of every month. He records the balance of each month in the graph above. During which one-month period did his balance decrease the most?
   (A) February-March
   (B) May-June
   (C) July-August
   (D) August-September
   (E) September-October

7. In December 2018, Leon decides to withdraw $50 for holiday gifts. At work, he receives a bonus check for $125 and deposits it. If no other deposits or withdrawals were made in December 2018, what is Leon's balance on January 1, 2019?
   (A) $300
   (B) $350
   (C) $400
   (D) $425
   (E) $475

8. The Khan family keeps a record of their expenses. The following graph shows their records. They save the rest of their income. If the Khans earned $80,000 in 2016, how much did they save that year?

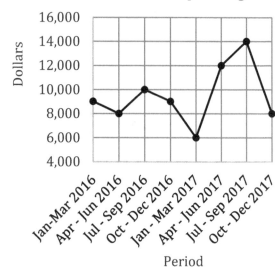

   (A) $4,000
   (B) $33,000
   (C) $36,000
   (D) $40,000
   (E) $44,000

Use the following graph to answer questions 9 and 10.

A team of scientists observed the growth of Plant X for 8 years. At the end of every year, they determined how much it rained and how much Plant X grew.

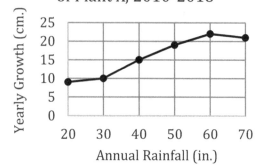

9. Which statement best describes the effects of rainfall on Plant X's growth?
   (A) Increasing rainfall always causes the plant to grow more rapidly.
   (B) Yearly growth increases until rainfall is 60 inches, then decreases rapidly.
   (C) Yearly growth increases until rainfall is 60 inches, then decreases slightly.
   (D) Rainfall less than 40 inches has no effect on growth.
   (E) Increasing rainfall from 40 to 50 inches affects growth more than increasing rainfall from 30 to 40 inches.

10. Based on the data, about how much rainfall is needed for Plant X to grow 12 cm?
    (A) 25 in.
    (B) 30 in.
    (C) 35 in.
    (D) 40 in.
    (E) 45 in.

Interpreting Circle Graphs

Use the following information and graph to answer questions 1 and 2.

Sixth graders at Eastside Park Middle School were asked what their favorite subject is. The circle graph below displays the results.

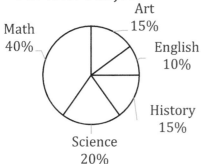

1. If 350 students were surveyed, how many said Math was their favorite subject?
   (A) 40
   (B) 70
   (C) 140
   (D) 400
   (E) 1400

2. Which two subjects did a total of one-quarter of the students surveyed say was their favorite?
   (A) Art and Math
   (B) Art and Science
   (C) Art and History
   (D) Science and History
   (E) History and English

Use the following information and graph to answer questions 3 and 4.

Every month, Olive receives an allowance of $40. Below is a circle graph displaying how she spends her money in a typical month.

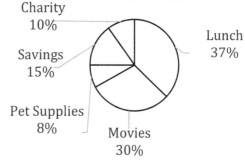

3. On which category does Olive typically spend $3.20 a month?
   (A) Movies
   (B) Pet Supplies
   (C) Savings
   (D) Charity
   (E) Lunch

4. Using the data for a typical month, determine how much of her annual allowance Olive saves in one year.
   (A) $6.00
   (B) $15.00
   (C) $72.00
   (D) $180.00
   (E) $312.00

Use the following information and graph to answer questions 5 and 6.

For math class, Angel was asked to conduct a survey and present his results in a pie chart. He decided to ask people at the grocery store what their favorite fruit is. His results are below.

Favorite Fruit

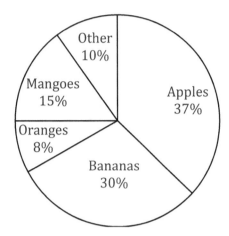

5. During his presentation, Angel was asked by a classmate how many people he surveyed. He couldn't remember the exact number off the top of his head, but he did remember that 12 people replied mangoes. Which mathematical expression could be evaluated to find the total number of people Angel surveyed?
   (A) 100 ÷ 12
   (B) 100 × 12
   (C) 12 × 15
   (D) 12 × 0.15
   (E) 12 ÷ 0.15

6. If 16 people said oranges were their favorite fruit, how many people said bananas were their favorite?
   (A) 8
   (B) 10
   (C) 18
   (D) 60
   (E) 80

Use the following information and graph to answer questions 7 and 8.

The owner of Pizza Deluxe decided to keep a tally of the toppings ordered on 200 pizzas. The chart below displays her data. Using this data, the owner decides to construct a circle graph for her display.

Toppings	Frequency
Pepperoni	68
Sausage	24
Veggies	38
Plain Cheese	50
Other	20

7. The slice representing "Other" toppings would have a central angle measuring how many degrees? *(Note: There are 360 degrees in a circle.)*
   (A) 10
   (B) 20
   (C) 36
   (D) 72
   (E) 144

8. For which topping would she use a central angle of 90 degrees?
   (A) Pepperoni
   (B) Sausage
   (C) Veggies
   (D) Plain Cheese
   (E) Other Toppings

Use the following information and graph to answer questions 9 and 10.

Every spring, the Drama Club at Franklin Middle School puts on a musical. Their expenses are displayed in the circle graph below.

9. If it costs $2,000 to put on a show, approximately how much money is spent on Set & Props?
   (A) $1,000
   (B) $750
   (C) $500
   (D) $50
   (E) $40

10. If the drama club spent $400 on Play Rights, approximately how much did they spend on Costumes?
    (A) $50
    (B) $100
    (C) $250
    (D) $450
    (E) $800

Statistics & Probability

Basic Probability

1. What is the probability of randomly selecting a triangle from the group of shapes below?

   (A) $\frac{1}{5}$
   (B) $\frac{1}{4}$
   (C) $\frac{1}{3}$
   (D) $\frac{1}{2}$
   (E) $\frac{1}{1}$

2. Priya is at a party that has three cheese pizzas, two pepperoni pizzas, and one supreme pizza. If she chooses a slice at random, what is the probability it will be cheese?
   (A) $\frac{1}{6}$
   (B) $\frac{1}{4}$
   (C) $\frac{1}{3}$
   (D) $\frac{1}{2}$
   (E) $\frac{3}{3}$

3. The probability of choosing a red marble from a bag of marbles is greater than $\frac{1}{3}$ and less than $\frac{2}{3}$. If there are 27 marbles in the bag, what is the greatest number of red marbles there could be?
   (A) 9
   (B) 13
   (C) 17
   (D) 18
   (E) 19

4. An aquarium at the Bakersfield Children's Museum as 21 sea-creatures: three sharks, two stingrays, four eels, five jellyfish, and the remaining are goldfish. If a creature is selected at random, what is the probability it is a goldfish?
   (A) $\frac{1}{3}$
   (B) $\frac{1}{2}$
   (C) $\frac{7}{20}$
   (D) $\frac{8}{21}$
   (E) $\frac{2}{3}$

5. A spinner is made up of 12 equal slices with numbers as shown below. What is the probability of spinning a number greater than 2?

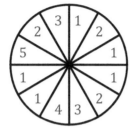

(A) $\frac{1}{6}$
(B) $\frac{1}{4}$
(C) $\frac{1}{3}$
(D) $\frac{3}{11}$
(E) $\frac{7}{12}$

6. A dartboard is shown below. The areas of regions 1, 2, 3, and 4 are $\frac{1}{10}, \frac{1}{5}, \frac{3}{10}$, and $\frac{2}{5}$ respectively. Hitting region 1 wins the grand prize, and hitting region 2 wins a small prize. What is the probability of winning either a grand prize or a small prize?

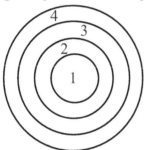

(A) $\frac{2}{15}$
(B) $\frac{3}{10}$
(C) $\frac{2}{5}$
(D) $\frac{1}{2}$
(E) $\frac{1}{3}$

7. What is the probability of randomly selecting a square from the group of shapes below?

□ △ ○ △ ☾ □ ○
○ □ □ ☆ ○ △ ○

(A) $\frac{1}{5}$
(B) $\frac{3}{14}$
(C) $\frac{2}{7}$
(D) $\frac{1}{3}$
(E) $\frac{5}{14}$

8. Victor will select one of the shapes below at random.

○ ○ △ □ ☆ ☆
○ △ ○ □ □ ☾

How many times greater is the probability of selecting a circle than that of selecting a triangle?

(A) 0
(B) $\frac{1}{2}$
(C) 1
(D) 2
(E) 3

9. Jaime's piggy bank contains pennies, nickels, dimes, and quarters. If he randomly selects a coin, what is the probability that it will be a dime?

Coins in Jaime's Piggy Bank

Coin Type	Number of Coins
Penny	23
Nickel	7
Dime	4
Quarter	6

(A) $\frac{1}{10}$
(B) $\frac{1}{9}$
(C) $\frac{3}{20}$
(D) $\frac{1}{4}$
(E) $\frac{1}{3}$

10. There is a parking lot full of cars with license plates from different states as shown in the chart below. If a car is chosen at random from the lot, what is the probability the car is NOT from California or New York?

State	Cars
California	13
New York	8
Texas	7
Other	7

(A) $\frac{1}{5}$
(B) $\frac{2}{7}$
(C) $\frac{1}{3}$
(D) $\frac{2}{5}$
(E) $\frac{3}{5}$

11. A company surveyed 60 employees to determine how their employees commute to work. The probabilities are shown in the chart below.

Method	Probability
Drive	$\frac{2}{5}$
Public Transportation	$\frac{1}{3}$
Walk	$\frac{1}{6}$
Bike	$\frac{1}{10}$

Which method of transportation do 20 of the employees surveyed use to commute to work?
(A) Driving
(B) Public Transportation
(C) Walk
(D) Bike
(E) Cannot be determined

12. A square dartboard is shown below. One point is rewarded for hitting the grey outer-ring; three points are rewarded for hitting in the white inner-ring, and ten points are rewarded for hitting the black bulls-eye. If one dart is thrown and hits the board, what is the probability of getting exactly three points? (Assume each gridded-square has the same area.)

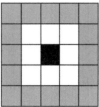

(A) $\frac{3}{25}$
(B) $\frac{1}{8}$
(C) $\frac{8}{25}$
(D) $\frac{9}{25}$
(E) $\frac{8}{17}$

13. A box contains nine chocolate candies, four licorice candies, six strawberry candies, one orange candy, and ten mints. Which candy has a $\frac{1}{5}$ chance of being selected?
(A) chocolate
(B) licorice
(C) orange
(D) mint
(E) strawberry

Compound Events

1. A fair coin was tossed 4 times and landed tails up all 4 times. What is the probability that the next toss will also result in tails?
   (A) 1
   (B) $\frac{3}{4}$
   (C) $\frac{1}{2}$
   (D) $\frac{1}{16}$
   (E) $\frac{1}{32}$

2. If Zoe randomly chooses one letter from the word MATH and one letter from the word ENGLISH, what is the probability that they are the same letter?
   (A) $\frac{1}{28}$
   (B) $\frac{1}{18}$
   (C) $\frac{1}{14}$
   (D) $\frac{1}{11}$
   (E) $\frac{2}{11}$

3. A number cube with the faces numbered 1 through 6 is rolled three times. What is the probability that all three rolls will result in the number one?
   (A) $\frac{1}{2}$
   (B) $\frac{1}{6}$
   (C) $\frac{3}{216}$
   (D) $\frac{1}{120}$
   (E) $\frac{1}{216}$

4. License plates issued in a certain county consist of 3 digits, each from 0-9. If each plate is assigned randomly, what is the chance a person gets a license plate that reads 888?
   (A) $\frac{1}{1,000}$
   (B) $\frac{3}{1,000}$
   (C) $\frac{1}{10}$
   (D) $\frac{3}{10}$
   (E) $\frac{4}{5}$

5. Ziggy has a box of stickers that are either blue, green, or yellow. There are an equal number of blue, green, and yellow stickers. Ziggy chooses a sticker at random, puts it back in the box, and then chooses a second sticker. What is the probability that Ziggy will choose a yellow sticker and then a blue sticker?
   (A) $\frac{1}{9}$
   (B) $\frac{2}{9}$
   (C) $\frac{1}{3}$
   (D) $\frac{2}{3}$
   (E) 1

6. One box contains a red sticker, a green sticker, and a blue sticker. Another box contains a red pencil, a green pencil, and a blue pencil. Gus randomly selects one sticker and one pencil. What is the probability that the sticker and pencil are the same color?
   (A) $\frac{2}{3}$
   (B) $\frac{1}{2}$
   (C) $\frac{1}{3}$
   (D) $\frac{1}{6}$
   (E) $\frac{1}{9}$

7. A bag contains 6 red marbles, 2 green marbles, and 2 blue marbles. What is the probability of randomly selecting a green marble and then a red marble, if no marbles are put back in the bag?
   (A) $\frac{3}{25}$
   (B) $\frac{2}{15}$
   (C) $\frac{2}{5}$
   (D) $\frac{8}{19}$
   (E) $\frac{4}{5}$

8. Sophie has a stack of 15 cards, numbered 1 through 15. If Sophie randomly selects two cards without replacing them, what is the probability that the numbers on both cards are divisible by five?
   (A) $\frac{1}{5}$
   (B) $\frac{2}{15}$
   (C) $\frac{1}{15}$
   (D) $\frac{1}{25}$
   (E) $\frac{1}{35}$

9. A box contains chocolates, each wrapped with a wrapper that is either green, blue, red, or purple. There are 2 chocolates with green wrappers, 3 with blue wrappers, 4 with red wrappers, and 1 with a purple wrapper. What is the probability of randomly selecting chocolate with a green wrapper and then selecting a chocolate with a purple wrapper, if no chocolates are replaced?
   (A) $\frac{3}{19}$
   (B) $\frac{3}{20}$
   (C) $\frac{1}{30}$
   (D) $\frac{1}{45}$
   (E) $\frac{1}{50}$

10. A box contains 10 yellow marbles, 7 blue marbles, and 3 green marbles. Valerie randomly selects one marble from the box, sets it aside, and then selects a second marble. What is the probability that Valerie will randomly select a yellow marble and then a green marble?
    (A) $\frac{3}{40}$
    (B) $\frac{3}{38}$
    (C) $\frac{4}{21}$
    (D) $\frac{13}{39}$
    (E) $\frac{13}{40}$

11. A bag contains 5 black cards, 5 red cards, and 5 white cards. Kai randomly selects 3 cards from the bag without replacing them. What is the probability that Kai will select 3 red cards?
    (A) 1
    (B) $\frac{1}{3}$
    (C) $\frac{1}{27}$
    (D) $\frac{2}{91}$
    (E) $\frac{4}{225}$

Mean, Median, Mode

1. Which set has the mode with the smallest value?
   (A) 1, 2, 8, 4, 8, 3, 7, 8
   (B) 13, 7, 9, 7, 15, 11, 12, 10
   (C) 0, 1, 2, 9, 8, 7, 9, 3
   (D) 7, 10, 3, 2, 10, 11, 15, 9
   (E) 3, 6, 9, 8, 6, 7, 6, 10

2. The list shows Malcolm's test scores. What is Malcolm's mean test score?
   80, 70, 100, 90, 80, 90
   (A) 75
   (B) 80
   (C) 85
   (D) 90
   (E) 100

3. Mark collects sports cards. Which sports card represents the median number of cards he's collected?

Sport	Number of Cards
Baseball	44
Football	36
Soccer	28
Basketball	34
Hockey	18

   (A) Hockey
   (B) Basketball
   (C) Football
   (D) Soccer
   (E) Baseball

4. The graph below shows the number of students who took part in each extracurricular activity at Washington Middle School one school year.

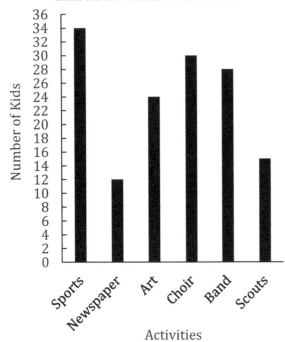

What is the range of students who participate in extracurricular activities displayed in the graph?
(A) 15
(B) 19
(C) 20
(D) 22
(E) 26

5. Jamila had six babysitting jobs in one month. She earned the following money from each job: $40, $60, $30, $75, $60, $36. What is the median amount Jamila earned from babysitting?
(A) 30
(B) 40
(C) 50
(D) 60
(E) 75

6. Students are selling rolls of wrapping paper to raise money for the school play. The chart shows the number of rolls of wrapping paper each student sold in one week.

Student	Number of Rolls
Sam	44
Priya	56
Mei	?
Michelle	52
Ricky	48

If the median number of tickets sold was 52, how many tickets could Mei have sold?
(A) 44
(B) 45
(C) 48
(D) 50
(E) 64

7. Regina had 5 books on a shelf. The chart below shows the number of pages in each book.

Book	Number of Pages
Book 1	256
Book 2	152
Book 3	130
Book 4	314
Book 5	282

Based on this list, which book has the median number of pages?
(A) Book 1
(B) Book 2
(C) Book 3
(D) Book 4
(E) Book 5

8. A real estate agent helped sell 5 houses with these prices:
    $155,000 $112,000 $170,000
       $220,000 $185,000
    What is the range of house prices?
    (A) 108,000
    (B) 112,000
    (C) 168,400
    (D) 170,000
    (E) 220,000

9. Stephanie shipped three boxes. The weight of the first box was 18 pounds. The second box weighed 12 pounds. If the mean of Stephanie's three boxes is 20, what was the weight of the third box?
    (A) 20
    (B) 30
    (C) 40
    (D) 50
    (E) 60

10. 10A baseball team scored the following number of runs during their first 9 games of the season: 5, 4, 4, 1, 5, 6, 3, 4, 2 If they want to end their season with a mean of 4 runs per game, how many runs must they score in their 10th (final) game?
    (A) 3
    (B) 4
    (C) 5
    (D) 6
    (E) 7

11. A swimmer tracked how long his laps took.

Lap	Time (Minutes)
1	0.9
2	1
3	1.4
4	?
5	1.6
6	1.2

    The mean time for all laps was 1.2 seconds. What was the time for lap 4?
    (A) 1.0
    (B) 1.1
    (C) 1.2
    (D) 1.3
    (E) 1.4

12. Angela has scored a total of 65 points in her first five basketball games. If she wants to increase her average score by 2 points, how many points does she have to score in her next game?
    (A) 11
    (B) 13
    (C) 15
    (D) 25
    (E) 67

13. Nine students in Mr. Brown's class took a quiz. Their scores were 88, 76, 92, 79, 88, 85, 99, 78, and 80. What is the difference between the mode and the median scores?
    (A) 0
    (B) 3
    (C) 9
    (D) 11.5
    (E) 23

Verbal

Overview

The Verbal section is comprised of synonyms and analogies. Synonym questions assess a student's ability to recognize words and reason through different relationships and subtle differences among words. The analogies measure how well students can relate words and ideas to each other using logic. Together, these questions are designed to test students' vocabulary and reasoning skills.

Students will have 60 minutes to answer 30 synonyms and 30 analogy questions.

On the Actual Test

Synonym questions consist of a single word in capital letters, followed by five answer choices labeled A through E. Students must select the answer choice that has the same or most nearly the same meaning as the word in capital letters.

Analogy questions ask students to look at a pair of words and select another pair of words with a matching relationship. Analogy questions are presented on the Middle Level SSAT in two different ways:

- Two-part stem: "A" is to "B" as…

  In these types of questions, students are given two words in the stem of the question, and are asked to select the pair of words with the same relationship as demonstrated between words "A" and "B."

- Three-part stem: "A" is to "B" as "C" is to…

  In these types of questions, students are given three words in the stem of the question; words "A" and "B" form one relationship, and are asked to complete the same relationship between word "C" and your answer choice.

There are many different types of relationships featured in the analogies. The most common ones are described below:

- Antonym relationships: Words that have the opposite, or nearly the opposite, meaning.
- Association relationships: One word in a pair is connected to the other word of the pair by association.
- Cause-and-Effect relationships: One word in a pair is a cause that may produce the other word.
- Defining relationships: Words that are related by a verb or verbal relationship.
- Degree/Intensity relationship: One word in a pair is related to the other word of the pair by a higher or lower degree.
- Function relationships: Words that define the function of an object.
- Grammatical relationships: Words that are related by a grammatical rule (e.g. plurality, tense, person, etc.).
- Individual to Object relationships: One word in a pair indicates a person or type of person, while the other word in the pair indicates the object which that person uses.
- Noun/Verb relationships: One word in a pair indicates a noun, while the other word of the pair indicates an action associated with that noun.
- Part/Whole relationships: The first word in a pair is a part of the second word of the pair.
- Purpose relationships: Words that have a common purpose when used together in a task.
- Type/Kind relationships: One word in a pair that is a type or kind of the other word of the pair.
- Whole/Part relationships: The first word in a pair is the whole, of which the second word is a part.

- Synonym relationships: Words that have very similar meaning (NOTE: not the same as the questions featured in the synonyms portion of the Verbal section).

Unlike in this workbook, the questions on the actual test will **not** indicate which type of relationship each analogy represents.

In This Practice Book

Synonym questions are given their own section and are generally presented in order of difficulty (the harder questions appear toward the end of the section).

Analogy questions are also given their own section and are also generally presented in order from less to more challenging. However, analogy questions are further subdivided as follows:

- Guided Practice – 14 units, one each for the individual types of analogies listed above. Each unit will only include a specific type of analogy. This allows students to focus on a particular type of analogy, if needed.
- Mixed Practice – 10 units of mixed practice. Each unit will include a mix of analogy types, which mirrors the format of the actual test. This allows students to see how well they know the different types of analogies.

There are additional instructions and recommendations at the beginning of each of the Synonyms and Analogies sections, which students should review before starting.

There are many ways to tackle this section of the test. Use the results from the diagnostic practice test to develop a plan. For instance, students may want to try the analogies Mixed Practice 1 to get a sense for which areas to focus on. Then, dive into the relevant Guided Practice sections. If students already know what need further work, they may jump right in to a Guided Practice section.

Tutorverse Tips!

Knowing whether or not to guess can be tricky. The Middle Level SSAT gives students 1 point for each question that is answered correctly. However, if students answer a question incorrectly, ¼ of a point will be deducted from the total score (see example below). No points will be awarded or deducted for questions that are left unanswered. Therefore, answer easy questions first, and come back to tougher questions later.

The formula for determining the raw score is:
(Number of Questions Answered Correctly × 1) – (Number of Questions Answered Incorrectly × ¼)

How Guessing Impacts A Score

Since there are 150 scored questions (167 total questions – 1 writing prompt – 16 experimental questions), the highest possible raw score is 150 points. This would be awarded to students who answer all 150 questions correctly.

If a student answers 110 questions correctly **but answers 40 questions incorrectly**, he or she will have earned (110 × 1) – (40 × ¼), or 110 – 10 = 100 points.

If a student answers 110 questions correctly **but leaves 40 questions unanswered**, he or she will have earned (110 × 1) – (40 × 0), or 110 – 0 = 110 points.

Therefore, it is important to refrain from making wild guesses. Instead, try to use process of elimination to make an educated guess.

Synonyms

Overview

Synonym questions consist of a single word in capital letters, followed by five answer choices labeled A through E. Students must select the answer choice that most nearly has the same meaning as the word in capital letters.

How to Use This Section

How much time students should spend on this section should be based on the diagnostic practice test results as well as the student's study plan. For most students, even those who score very well on the diagnostic practice test, we recommend practicing at least 10-15 questions per week in preparing for the exam. Those who score well on the diagnostic practice test and have an expansive vocabulary may wish to focus more on intermediate and advanced questions, while other students may wish to focus on introductory and intermediate questions.

The purpose of this section is to introduce students to new words. Some may find many of the words in this section to be challenging. Students should not be surprised to have to look up many of the words encountered in this section! We encourage students to make a list of difficult or challenging words, whether they appear in questions or answer choices. Write down the definition of each word as well as a sentence using the word. Students might also want to consider writing down positive or negative associations, any root words that can help them remember the word, or any words that are commonly encountered with that word.

Tutorverse Tips!

Sometimes, words can have more than one meaning. Don't be confused! Look at the answer choices to make an educated guess as to which meaning is being used in the question. Then, use reasoning skills to select the word that most nearly means the same as the word in capital letters.

Use it in a Sentence

As you read the question words in capital letters, think of a word that you might use instead of the question word.

If you don't know what a word means, try using it in a sentence. This will often help you see which word can be used to replace the question word. If this doesn't help, see if you can figure out whether or not the word has any positive or negative associations that match those of the answer choices.

Study Roots, Prefixes, & Suffixes

Many English words are derived from a single Greek or Latin root word. Sometimes, these root words relate to many different English words. In addition, knowing common prefixes and suffixes can help with very long or unfamiliar words, which can inform on the word's connotation or meaning.

Read, Read, Read!

Finally, there is no better preparation for the Synonym section than spending time reading. The practice of reading, whether for school or for pleasure, will help you build up your vocabulary. It will give you practice in utilizing context clues and figuring out what unknown words might mean. Reading at or above your current grade level will help you make better sense of more complicated words and sentences.

Introductory

Directions – Select the one word or phrase whose meaning is closest to the word in capital letters.

1. ABANDON:
   (A) run
   (B) subterranean
   (C) display
   (D) leave
   (E) defend

2. REBELLIOUS:
   (A) tame
   (B) punish
   (C) convenient
   (D) outrage
   (E) disobedient

3. RECOVER:
   (A) cherish
   (B) enthusiastic
   (C) blanket
   (D) wellness
   (E) heal

4. REHABILITATION:
   (A) recovery
   (B) vacation
   (C) building
   (D) obvious
   (E) formula

5. RESEARCH:
   (A) assistant
   (B) textbook
   (C) substitute
   (D) study
   (E) consultation

6. REVOLUTION:
   (A) amplification
   (B) revolt
   (C) army
   (D) calm
   (E) wheel

7. EDIT:
   (A) delete
   (B) type
   (C) journal
   (D) change
   (E) paper

8. SITE:
   (A) quote
   (B) detest
   (C) concise
   (D) location
   (E) stand

9. SOMEWHAT:
   (A) amendment
   (B) free
   (C) kindly
   (D) casual
   (E) slightly

10. STRESS:
    (A) parallel
    (B) pressure
    (C) docile
    (D) environment
    (E) test

11. STYLE:
    (A) apologize
    (B) fashion
    (C) art
    (D) conquest
    (E) bountiful

12. SUBMISSION:
    (A) prize
    (B) notion
    (C) professional
    (D) participant
    (E) entry

13. TEAM:
    (A) crew
    (B) cheer
    (C) join
    (D) audition
    (E) game

14. UNATTRACTIVE:
    (A) astonishing
    (B) completely
    (C) forced
    (D) disgust
    (E) ugly

15. UNFINISHED:
    (A) incomplete
    (B) left
    (C) assert
    (D) pertain
    (E) ending

16. UNWORTHY:
    (A) next to
    (B) more value
    (C) not deserving
    (D) above all
    (E) fun filled

17. WIDESPREAD:
    (A) not common
    (B) all over
    (C) just enough
    (D) sharp pain
    (E) long story

18. INDEX:
    (A) finger
    (B) agency
    (C) list
    (D) series
    (E) second

19. INSTANCE:
    (A) section
    (B) external
    (C) status
    (D) occasion
    (E) project

20. JOB:
    (A) working
    (B) rules
    (C) prejudice
    (D) boss
    (E) occupation

21. MINOR:
    (A) adult
    (B) more
    (C) decrease
    (D) unimportant
    (E) excessive

22. NEUTRAL:
    (A) bright
    (B) natural
    (C) unbiased
    (D) favorite
    (E) mental

23. OVERALL:
    (A) generally
    (B) above
    (C) within
    (D) seldom
    (E) outside

24. PARTNER:
    (A) position
    (B) applicant
    (C) gang
    (D) company
    (E) ally

25. DISPOSABLE:
    (A) expendable
    (B) affirm
    (C) income
    (D) durable
    (E) garbage

26. PROCESS:
    (A) redo
    (B) parade
    (C) funeral
    (D) perform
    (E) method

27. EQUIP:
    (A) weapon
    (B) joke
    (C) prop
    (D) supply
    (E) tool

28. PRESUME:
    (A) initial
    (B) repel
    (C) preview
    (D) suppose
    (E) betray

29. DISGRACE:
    (A) graceful
    (B) erratic
    (C) negatively
    (D) shame
    (E) embarrassed

30. EXCEL:
    (A) achieve
    (B) calculate
    (C) compete
    (D) isolate
    (E) guarantee

31. IDENTIFY:
    (A) character
    (B) clue
    (C) detect
    (D) mask
    (E) contradict

32. ENVIRONMENT:
    (A) biological
    (B) current
    (C) natural
    (D) estate
    (E) surroundings

33. CATASTROPHE:
    (A) anniversary
    (B) disaster
    (C) event
    (D) happening
    (E) reminder

Intermediate

Directions – Select the one word or phrase whose meaning is closest to the word in capital letters.

1. ABUNDANCE:
    (A) be sick of
    (B) high risk
    (C) turn around
    (D) great amount
    (E) very wealthy

2. PHASE:
    (A) stage
    (B) contract
    (C) stall
    (D) admit
    (E) project

3. PREDOMINANT:
    (A) powerfully
    (B) acceptable
    (C) primary
    (D) random
    (E) vary

4. ADVENTUROUS:
    (A) deadly
    (B) acquaintance
    (C) excursion
    (D) anxious
    (E) outgoing

5. DESOLATE:
    (A) erase
    (B) tardy
    (C) empty
    (D) alien
    (E) wasteful

6. PROMOTE:
    (A) benevolent
    (B) advertise
    (C) announce
    (D) enlarge
    (E) television

7. PROPOSAL:
    (A) basis
    (B) wedding
    (C) suggestion
    (D) marry
    (E) basic

8. RELY:
    (A) befriend
    (B) prevail
    (C) sleep
    (D) depend
    (E) liar

9. ROLE:
    (A) acting
    (B) selfish
    (C) function
    (D) model
    (E) challenge

10. AWARE:
    (A) watch
    (B) decline
    (C) conscious
    (D) flexible
    (E) noticeable

11. STATISTIC:
    (A) confidential
    (B) state
    (C) factual
    (D) data
    (E) catalog

12. COMMUNICATION:
    (A) computer
    (B) message
    (C) device
    (D) accident
    (E) telephone

13. DEVOTE:
    (A) church
    (B) prayer
    (C) send
    (D) dedicate
    (E) alternate

14. AREA:
    (A) void
    (B) amateur
    (C) farmland
    (D) space
    (E) acre

15. VISION:
    (A) leader
    (B) glasses
    (C) eyeball
    (D) selfless
    (E) sight

16. ADVERSITY:
    (A) enemy
    (B) advertise
    (C) diversity
    (D) challenging
    (E) misfortune

17. DESIGN:
    (A) school
    (B) artist
    (C) clothing
    (D) plan
    (E) arcade

18. ANALYZE:
    (A) agitate
    (B) examine
    (C) artificial
    (D) calculation
    (E) evaluation

19. MINIMIZE:
    (A) counterfeit
    (B) miniature
    (C) reveal
    (D) claim
    (E) lessen

20. GLOBE:
    (A) world
    (B) round
    (C) historical
    (D) international
    (E) theater

21. GRADE:
    (A) letter
    (B) rank
    (C) student
    (D) teacher
    (E) test

22. LEGAL:
    (A) lawyer
    (B) allowed
    (C) careful
    (D) stable
    (E) court

23. ASPECT:
    (A) figurative
    (B) trait
    (C) structure
    (D) fact
    (E) absent

24. HENCE:
    (A) far
    (B) around
    (C) fence
    (D) therefore
    (E) behind

25. IMPRACTICAL:
    (A) joker
    (B) useful
    (C) admire
    (D) unrealistic
    (E) ambiguous

26. INDEFINITE:
    (A) invisible
    (B) detained
    (C) detached
    (D) unspecific
    (E) decisive

27. ELIMINATE:
    (A) remove
    (B) aggravate
    (C) creation
    (D) ponder
    (E) arrogant

28. INTERNAL:
    (A) inner
    (B) hide
    (C) brief
    (D) consist
    (E) secret

29. INVESTIGATE:
    (A) examine
    (B) clue
    (C) specific
    (D) detective
    (E) invest

30. APPROXIMATE:
    (A) certain
    (B) set
    (C) inexact
    (D) precise
    (E) demanding

31. RIGID:
    (A) stiff
    (B) simple
    (C) elevated
    (D) effective
    (E) irresponsible

32. BELITTLE:
    (A) small
    (B) criticize
    (C) depress
    (D) provoke
    (E) shrink

33. ENDURE:
    (A) last
    (B) length
    (C) strength
    (D) training
    (E) workout

Advanced

Directions – Select the one word or phrase whose meaning is closest to the word in capital letters.

1. PRECAUTION:
   - (A) prevent
   - (B) safeguard
   - (C) provide
   - (D) ignore
   - (E) warn

2. PRINCIPLE:
   - (A) error
   - (B) concept
   - (C) fantastic
   - (D) royalty
   - (E) teacher

3. PROJECTION:
   - (A) television
   - (B) screen
   - (C) bulletin
   - (D) prediction
   - (E) devious

4. RANGE:
   - (A) length
   - (B) dense
   - (C) average
   - (D) reverse
   - (E) portion

5. RECOGNITION:
   - (A) appreciation
   - (B) correct
   - (C) assignment
   - (D) bountiful
   - (E) knowledge

6. RESPIRATION:
   - (A) exercise
   - (B) breathing
   - (C) exception
   - (D) available
   - (E) sweat

7. AUGMENT:
   - (A) dig
   - (B) boost
   - (C) quarrel
   - (D) reduce
   - (E) require

8. SUBSTANCE:
   - (A) rotate
   - (B) lavish
   - (C) material
   - (D) substantial
   - (E) greedy

9. SUPPRESS:
   - (A) exhibit
   - (B) touch
   - (C) smother
   - (D) notify
   - (E) manual

10. COMMITTEE:
    - (A) deal
    - (B) chair
    - (C) progressive
    - (D) commit
    - (E) group

11. THEREBY:
    - (A) before
    - (B) thus
    - (C) always
    - (D) between
    - (E) beyond

12. TRACE:
    - (A) amount
    - (B) linger
    - (C) catastrophic
    - (D) run
    - (E) follow

13. NEVERTHELESS:
    - (A) unless
    - (B) however
    - (C) rightfully
    - (D) including
    - (E) wherever

14. ABBREVIATE:
    - (A) uniform
    - (B) shorten
    - (C) briefly
    - (D) fewer
    - (E) reveal

15. ABSOLUTE:
    - (A) finite
    - (B) curious
    - (C) certain
    - (D) comfort
    - (E) end

16. ODD:
    - (A) strangely
    - (B) unusual
    - (C) benign
    - (D) number
    - (E) annoying

17. ABSTRACT:
    - (A) theoretical
    - (B) art
    - (C) impact
    - (D) general
    - (E) hesitant

18. FINAL:
    - (A) consume
    - (B) last
    - (C) pitch
    - (D) civil
    - (E) conclude

19. EXOTIC:
    - (A) toxic
    - (B) contained
    - (C) wilderness
    - (D) foreign
    - (E) food

20. FOCUS:
    - (A) lens
    - (B) precise
    - (C) issue
    - (D) spotlight
    - (E) intensity

21. FUND:
    - (A) entertaining
    - (B) expensive
    - (C) mutual
    - (D) dollar
    - (E) provide

22. OBJECTIVE:
    - (A) league
    - (B) assort
    - (C) item
    - (D) goal
    - (E) variable

23. POLICY:
    - (A) police
    - (B) original
    - (C) safety
    - (D) political
    - (E) rule

24. ADJOURNED:
    - (A) concluded
    - (B) judged
    - (C) meeting
    - (D) coarse
    - (E) planned

25. INSATIABLE:
    (A) uncontrollable
    (B) insulting
    (C) consecutive
    (D) inconsiderate
    (E) hunger

26. MAJOR:
    (A) main
    (B) close
    (C) greatly
    (D) general
    (E) soldier

27. ELEMENT:
    (A) contribute
    (B) temporary
    (C) oxygen
    (D) water
    (E) component

28. MEDICINAL:
    (A) hospital
    (B) headache
    (C) doctor
    (D) healing
    (E) harmful

29. ADVOCATE:
    (A) secret admirer
    (B) vocal supporter
    (C) sneaky thief
    (D) hateful traitor
    (E) powerful person

30. DELUSION:
    (A) fact
    (B) liar
    (C) elusive
    (D) magician
    (E) misconception

31. CEASE:
    (A) stop
    (B) celebrate
    (C) fold
    (D) contagious
    (E) conclusion

32. CHANNEL:
    (A) boat
    (B) television
    (C) sailing
    (D) pathway
    (E) coursework

33. CORE:
    (A) value
    (B) steady
    (C) muscle
    (D) strong
    (E) center

Analogies

Overview

Analogy questions ask students to look at a pair of words and select another pair of words with a matching relationship. Analogy questions are presented on the Middle Level SSAT in two different ways:

- Two-part stem: "A" is to "B" as…
  In these types of questions, students are given two words in the stem of the question, and are asked to select the pair of words with the same relationship as demonstrated between words "A" and "B."

- Three-part stem: "A" is to "B" as "C" is to…
  In these types of questions, students are given three words in the stem of the question; words "A" and "B" form one relationship, and are asked to complete the same relationship between word "C" and your answer choice.

For the different types of analogy relationships, see p. 98 in the Verbal Overview section.

How to Use This Section

How much time students spend on this section should be based on their diagnostic practice test results as well as their study plan. For most students, even those who score very well on the diagnostic practice test, we recommend practicing at least 5-10 questions per week in preparing for the exam.

One study plan might be to try a Mixed Practice section to get a sense for which areas to focus on. Based on those results, work on the relevant Guided Practice sections. If students already know what they need to work on, they can jump right in to a Guided Practice section.

The purpose of this section is to introduce students to new words. Some may find many of the words in this section to be challenging. Students should not be surprised to have to look up many of the words encountered in this section! We encourage students to make a list of difficult or challenging words, whether they appear in questions or answer choices. Write down the definition of each word as well as a sentence using the word. Students might also want to consider writing down positive or negative associations, any root words that can help them remember the word, or any words that are commonly encountered with that word.

Tutorverse Tips!

The key to success on the analogies section is practice. Students should know the common types of analogies by heart and be able to quickly categorize the question stem provided into one of these types.

Write down the relationship in the margin and see which of the choices creates an analogy that matches that of the question stem.

Verbal 111

Guided Practice

Directions: For each question, select the answer choice that best completes the meaning of the sentence.

Antonyms

1. Despicable is to lovable as
   (A) delicious is to tasty
   (B) distant is to close
   (C) liquid is to wet
   (D) table is to chair
   (E) nice is to polite

2. Peaceful is to rowdy as
   (A) tall is to grand
   (B) build is to hammer
   (C) festive is to playful
   (D) learn is to forget
   (E) charming is to naïve

3. Severe is to mild as
   (A) rug is to carpet
   (B) hot is to melting
   (C) destination is to road
   (D) talk is to speak
   (E) bond is to separate

4. Picturesque is to grotesque as
   (A) maintain is to agreement
   (B) vast is to spacious
   (C) computer is to desk
   (D) diverse is to identical
   (E) grass is to lawn

5. Anxiously is to calmly as
   (A) lazily is to energetically
   (B) daily is to weekly
   (C) fly is to voyage
   (D) machine is to oil
   (E) softly is to gently

6. Distinctive is to ordinary as
   (A) rarely is to peculiar
   (B) end is to conclude
   (C) cover is to lodge
   (D) never is to always
   (E) embrace is to murmur

7. Famous is to unknown as separate is to
   (A) equal
   (B) different
   (C) together
   (D) fence
   (E) tricky

8. Subsequent is to previous as obscure is to
   (A) common
   (B) hidden
   (C) bashful
   (D) argumentative
   (E) definition

Cause-and-Effect

1. Artisan is to product as
   (A) couch is to comfortable
   (B) headache is to aspirin
   (C) bathtub is to bath
   (D) bacteria is to infection
   (E) send is to email

2. Voice is to noise as
   (A) bee is to honey
   (B) cow is to mule
   (C) farm is to farmer
   (D) wasp is to stinger
   (E) rake is to shovel

3. Comedy is to laughter as
   (A) reunion is to family
   (B) feline is to pet
   (C) security is to danger
   (D) eraser is to pencil
   (E) radiator is to warmth

4. Farm is to crops as
   (A) misfortune is to fortune
   (B) signal is to gesture
   (C) radio is to sound
   (D) affection is to output
   (E) sky is to antenna

The Tutorverse
www.thetutorverse.com

5. Labor is to fatigue as
   (A) glass is to clear
   (B) pavement is to cement
   (C) high is to low
   (D) projector is to image
   (E) hatred is to kindness

6. Apology is to forgiveness as
   (A) floral is to garden
   (B) diary is to journal
   (C) phone is to tone
   (D) fictitious is to real
   (E) happiness is to smiling

7. Moisture is to humidity as stretching is to
   (A) rubber
   (B) gas
   (C) waist
   (D) band
   (E) flexibility

8. Loss is to grief as dissatisfaction is to
   (A) complaint
   (B) suggestion
   (C) satisfaction
   (D) agree
   (E) hunger

Defining

1. Planet is to sun as
   (A) lamp is to moth
   (B) star is to meteor
   (C) moon is to earth
   (D) Pluto is to Jupiter
   (E) lake is to sea

2. Donkey is to wagon as
   (A) horse is to carriage
   (B) scientist is to chemical
   (C) telephone is to call
   (D) hypnotic is to magic
   (E) literature is to poet

3. Arborist is to trees as
   (A) thought is to teacher
   (B) silver is to jeweler
   (C) wine is to beer
   (D) gardener is to rose
   (E) fisherman is to fish

4. Pirate is to treasure as
   (A) chest is to gold
   (B) loot is to steal
   (C) adventure is to journey
   (D) sailor is to land
   (E) lost is to wayward

5. Cat is to yarn as
   (A) baby is to toy
   (B) tiger is to puma
   (C) child is to sister
   (D) elevator is to lift
   (E) bottle is to plastic

6. Professor is to student as
   (A) board is to chalk
   (B) coach is to athlete
   (C) ray is to sunshine
   (D) igloo is to ice
   (E) sapling is to acorn

7. Water is to fire as beverage is to
   (A) hunger
   (B) food
   (C) thirst
   (D) juice
   (E) drink

8. Compass is to direction as
   (A) rebellion is to radical
   (B) forgot is to remembered
   (C) groundhog is to shadow
   (D) refrigerator is to chilled
   (E) clock is to time

Degree/Intensity

1. Toxic is to unhealthy as
   (A) smart is to brilliant
   (B) numerous is to countless
   (C) icy is to road
   (D) tired is to cranky
   (E) microscopic is to small

2. Dry is to parched as
   (A) secretly is to secret
   (B) desert is to oasis
   (C) damp is to drenched
   (D) helpful is to weak
   (E) motionless is to mobile

3. Unavoidable is to likely as
   (A) crowd is to crowded
   (B) lively is to affectionate
   (C) wounded is to suffering
   (D) gash is to scratch
   (E) final is to ended

4. Worship is to like as
   (A) bawl is to sniffle
   (B) cry is to tearful
   (C) idolize is to hero
   (D) fight is to combat
   (E) excited is to rested

5. Accident is to catastrophe as
   (A) poorly is to badly
   (B) mishap is to disaster
   (C) abnormal is to unusual
   (D) creation is to destruction
   (E) faulty is to confusion

6. Warm is to scald as
   (A) construction is to zone
   (B) boil is to cook
   (C) mix is to dilute
   (D) teach is to brainwash
   (E) shake is to stir

7. Genius is to smart as ancient is to
   (A) young
   (B) old
   (C) monument
   (D) recent
   (E) archaeologist

8. Despairing is to sad as enraged is to
   (A) blissful
   (B) confused
   (C) upset
   (D) discouraged
   (E) perfumed

Function/Object

1. Mask is to hide as
   (A) cane is to walk
   (B) wander is to safari
   (C) balance is to steady
   (D) problem is to issue
   (E) sacred is to holy

2. Army is to defend as
   (A) recycle is to cardboard
   (B) elevate is to platform
   (C) police is to protect
   (D) plea is to rejection
   (E) monk is to priest

3. Glasses is to see as
   (A) society is to trust
   (B) accountant is to count
   (C) ritual is to customary
   (D) spokesperson is to beg
   (E) rider is to jockey

4. Mail is to communicate as
   (A) cabin is to protection
   (B) feign is to fake
   (C) extension is to border
   (D) assistant is to aid
   (E) examine is to specimen

5. Pantry is to save as
   (A) vacuum is to tidy
   (B) operation is to define
   (C) viable is to necessary
   (D) outrage is to shock
   (E) biscuit is to butter

6. Synopsis is to summarize as
   (A) violation is to penalty
   (B) contribution is to withdrawal
   (C) concede is to lose
   (D) rival is to challenge
   (E) fund is to tax

7. Conqueror is to win as voter is to
   (A) candidacy
   (B) elect
   (C) district
   (D) slogan
   (E) ballot

8. Counselor is to advise as contract is to
   (A) agree
   (B) unrest
   (C) handshake
   (D) legal
   (E) signed

Grammar

1. Skip is to skipping as
   (A) confront is to confronted
   (B) eat is to eaten
   (C) taste is to tasting
   (D) sleeping is to sleep
   (E) expose is to exposed

2. Order is to ordered as
   (A) dinner is to lunch
   (B) command is to obey
   (C) dug is to dig
   (D) swim is to swam
   (E) jumping is to jumped

3. Catered is to cater as informed is to
   (A) caters
   (B) catering
   (C) informs
   (D) informing
   (E) inform

4. Am is to was as
   (A) gone is to gotten
   (B) go is to gone
   (C) is to are
   (D) go is to went
   (E) are is to be

5. Taught is to teach as
   (A) eat is to ate
   (B) packaged is to package
   (C) student is to school
   (D) learning is to wisdom
   (E) pass is to fail

6. Devastate is to devastated as contradict
   (A) contradiction
   (B) contradicting
   (C) contradicted
   (D) devastating
   (E) devastates

7. Foot is to feet as mouse is to
   (A) mouses
   (B) mice
   (C) moused
   (D) miced
   (E) mices

8. Knives is to knife as leaves is to
   (A) hone
   (B) sharpening
   (C) leaf
   (D) leaver
   (E) left

Individual/Object

1. Designer is to pen as
   - (A) teacher is to chalkboard
   - (B) museum is to architecture
   - (C) label is to identify
   - (D) moral is to fable
   - (E) activity is to pastime

2. Nurse is to thermometer as
   - (A) biologist is to microscope
   - (B) cure is to remedy
   - (C) data is to statistics
   - (D) marine is to wildlife
   - (E) assist is to comfort

3. Carpenter is to nail as
   - (A) warrant is to arrest
   - (B) sewing is to stitching
   - (C) tailor is to needle
   - (D) rooftop is to foundation
   - (E) delicate is to breakable

4. Baker is to oven as
   - (A) church is to pew
   - (B) kitchen is to blender
   - (C) ingredient is to scrumptious
   - (D) bucket is to cleaner
   - (E) chef is to stove

5. Driver is to vehicle as
   - (A) mechanic is to plumber
   - (B) cyclist is to bicycle
   - (C) wheel is to tire
   - (D) wrench is to rotate
   - (E) accident is to collision

6. Shopper is to money as
   - (A) consumed is to bought
   - (B) option is to decision
   - (C) judge is to gavel
   - (D) repellant is to roach
   - (E) pesticide is to exterminator

7. Pitcher is to baseball as catcher is to
   - (A) bat
   - (B) game
   - (C) glove
   - (D) umpire
   - (E) diamond

8. Firefighter is to hose as student is to
   - (A) robbery
   - (B) thievery
   - (C) stole
   - (D) jail
   - (E) pencil

Noun/Verb

1. Toast is to bread as
   - (A) unruly is to bombastic
   - (B) hibernate is to winter
   - (C) display is to trophy
   - (D) captor is to hostage
   - (E) warning is to telltale

2. Earn is to medal as
   - (A) command is to ambiguous
   - (B) scribble is to decipher
   - (C) unclear is to code
   - (D) door is to hinge
   - (E) wrap is to present

3. Follow is to path as
   - (A) street is to mile
   - (B) runaway is to homeless
   - (C) atmosphere is to vast
   - (D) keep is to secret
   - (E) costly is to damage

4. Apply is to job as
   - (A) gracious is to idyllic
   - (B) appraise is to friend
   - (C) rumor is to spread
   - (D) demolish is to technology
   - (E) make is to mistake

5. Extinguish is to flame as
   (A) cautiously is to warily
   (B) foolhardy is to rugged
   (C) corner is to sharp
   (D) mesmerized is to amused
   (E) breathe is to air

6. Nurture is to newborn as
   (A) justify is to encouragement
   (B) autograph is to signature
   (C) dash is to sprint
   (D) friend is to connection
   (E) operate is to machinery

7. Cut is to paper as praise is to
   (A) hero
   (B) prized
   (C) fool
   (D) candidate
   (E) best

8. Devour is to snack as enforce is to
   (A) rule
   (B) police
   (C) weapon
   (D) game
   (E) referee

Part/Whole

1. Stanza is to poem as
   (A) neck is to necklace
   (B) jar is to lid
   (C) letter is to alphabet
   (D) iron is to shirt
   (E) right is to left

2. Finger is to hand as
   (A) figure is to draw
   (B) screen is to automatic
   (C) brick is to wall
   (D) manual is to grasp
   (E) car is to bus

3. Branch is to tree as
   (A) flight is to airplane
   (B) slice is to pie
   (C) cooked is to baked
   (D) technician is to survey
   (E) consume is to dessert

4. Grain is to dune as
   (A) photograph is to develop
   (B) boulder is to beach
   (C) green is to blue
   (D) sheet is to pillow
   (E) pane is to window

5. Flock is to pigeon as
   (A) duck is to duckling
   (B) television is to internet
   (C) cream is to gel
   (D) building is to floor
   (E) solid is to material

6. Orchestra is to violinist as
   (A) price is to cost
   (B) flag is to pole
   (C) mall is to store
   (D) pleasant is to audible
   (E) east is to west

7. Soldier is to army as ant is to
   (A) anteater
   (B) colony
   (C) cohabitate
   (D) termite
   (E) hive

8. Star is to constellation as side is to
   (A) square
   (B) long
   (C) measurement
   (D) diagonal
   (E) flood

Purpose/Object

1. Tape is to papers as
   (A) adhere is to peel
   (B) recording is to speech
   (C) salmon is to stream
   (D) hook is to sink
   (E) nail is to boards

2. Period is to sentence as
   (A) dynamite is to explosion
   (B) punctuation is to mark
   (C) finale is to play
   (D) hunter is to farmer
   (E) dagger is to cleaver

3. Sponge is to dish as
   (A) electronic is to electricity
   (B) melody is to tune
   (C) landscape is to viewer
   (D) broom is to floor
   (E) mop is to soap

4. Rifle is to bullet as
   (A) disgrace is to shame
   (B) fertilizer is to dirt
   (C) weapon is to illegal
   (D) odor is to foul
   (E) harmful is to dangerous

5. Calculator is to numbers as
   (A) plate is to flat
   (B) scuba is to underwater
   (C) knife is to food
   (D) revive is to promote
   (E) thunder is to lightning

6. Chisel is to stone as
   (A) linked is to chained
   (B) hate is to loathe
   (C) pill is to swallow
   (D) chainsaw is to wood
   (E) conjunction is to and

7. Recipe is to meal as map is to
   (A) North
   (B) soldier
   (C) harbor
   (D) trip
   (E) compass

8. Lotion is to rash as bandage is to
   (A) ambulance
   (B) wound
   (C) itch
   (D) nurse
   (E) leg

Type/Kind

1. Color is to violet as
   (A) prop is to mechanism
   (B) demise is to death
   (C) order is to request
   (D) reflex is to stretch
   (E) metal is to titanium

2. Animal is to lion as
   (A) restaurant is to cafe
   (B) liquefy is to smoothie
   (C) barbarian is to savage
   (D) dispute is to refute
   (E) stack is to pile

3. Boat is to canoe as
   (A) shoe is to sandal
   (B) fluorescent is to radiate
   (C) consideration is to regret
   (D) candy is to flavor
   (E) collar is to button

4. Flower is to rose as
   (A) boot is to snow
   (B) field is to dugout
   (C) chess is to piece
   (D) ball is to basketball
   (E) healthy is to nutritious

5. Mineral is to emerald as
   (A) strength is to fortitude
   (B) sport is to football
   (C) transmit is to receive
   (D) normal is to original
   (E) fiction is to story

6. Artist is to sculptor as
   (A) vault is to storage
   (B) dilemma is to trauma
   (C) homemaker is to residence
   (D) curve is to bend
   (E) hobby is to knitting

7. Doctor is to pediatrician as fruit is to
   (A) plant
   (B) skin
   (C) apple
   (D) potato
   (E) seed

8. Interaction is to greeting as furniture is to
   (A) plush
   (B) sofa
   (C) leather
   (D) recline
   (E) stuffed

Synonym

1. Accidentally is to unexpectedly as
   (A) headphones is to sound
   (B) minor is to major
   (C) feeble is to age
   (D) marriage is to ceremony
   (E) sleepy is to drowsy

2. Envious is to jealous as
   (A) serious is to sane
   (B) slow is to turtle
   (C) skyscraper is to building
   (D) messenger is to message
   (E) enchanted is to charmed

3. Infamous is to notorious as
   (A) school is to spirit
   (B) carefully is to careful
   (C) evening is to dark
   (D) toxic is to poisonous
   (E) vacate is to travel

4. Malicious is to spiteful as
   (A) blazing is to scorching
   (B) young is to elderly
   (C) earn is to dollars
   (D) closet is to clothes
   (E) simple is to ornate

5. Stealthily is to sneakily as
   (A) promptly is to punctually
   (B) wearily is to tired
   (C) proprietary is to own
   (D) encompass is to navigate
   (E) marvel is to comics

6. Fragile is to dainty as
   (A) merely is to only
   (B) concur is to disagree
   (C) gentle is to violent
   (D) laxity is to strict
   (E) loudly is to volume

7. Timid is to shy as remote is to
   (A) television
   (B) past
   (C) far
   (D) relevant
   (E) central

8. Doubtful is to uncertain as suspicious is to
   (A) character
   (B) mystery
   (C) thief
   (D) skeptical
   (E) unquestionable

Mixed Practice

Directions: For each question, select the answer choice that best completes the meaning of the sentence.

1. Village is to metropolis as
   (A) human is to elephant
   (B) starving is to thirsty
   (C) deep is to bottomless
   (D) bear is to cub
   (E) anthill is to beehive

2. Shield is to defend as
   (A) mow is to meadow
   (B) basis is to formula
   (C) step is to stomp
   (D) chorus is to together
   (E) party is to celebrate

3. Tornado is to destruction as
   (A) politician is to history
   (B) pleased is to wealthy
   (C) medicine is to health
   (D) opera is to composer
   (E) alternate is to various

4. Rubric is to grades as
   (A) memo is to note
   (B) item is to catalog
   (C) law is to punishment
   (D) curtain is to rod
   (E) lunar is to solar

5. Researcher is to encyclopedia as
   (A) adversary is to enemy
   (B) cartographer is to map
   (C) bewitch is to confuse
   (D) substitute is to replacement
   (E) television is to broadcast

6. Illogical is to nonsensical as
   (A) secede is to join
   (B) impertinent is to important
   (C) conspire is to sweat
   (D) foretell is to anecdote
   (E) divert is to redirect

7. Concoct is to mixture as
   (A) educate is to pupil
   (B) criticize is to empathy
   (C) break is to ruin
   (D) compartment is to sealed
   (E) deserve is to patience

8. Vital is to unnecessary as
   (A) smarts is to intelligence
   (B) universe is to space
   (C) key is to essential
   (D) aged is to elderly
   (E) junior is to senior

9. T-shirt is to outfit as
   (A) cotton is to silk
   (B) rotation is to turn
   (C) wardrobe is to ensemble
   (D) layer is to cake
   (E) style is to flair

10. Graduate is to graduated as
    (A) ceremony is to gown
    (B) finish is to complete
    (C) celebrated is to celebrate
    (D) water is to watered
    (E) instructed is to instruct

11. Fish is to tuna as
    (A) tedious is to devout
    (B) habitat is to unnatural
    (C) sack is to carry
    (D) lure is to catch
    (E) reptile is to snake

12. Ballerina is to ballet as singer is to
    (A) orchestra
    (B) opera
    (C) pitch
    (D) voice
    (E) songwriter

Reading

Overview

In the Reading section, students will read passages and answer questions that pertain to those passages. The passages will vary; some passages will be short poems, while others will be longer essays. All of the questions are designed to measure how well a student understands what he or she reads.

Students will have 40 minutes to answer 40 reading comprehension questions.

On the Actual Test

There are two main types of passages on the Middle Level SSAT: fiction and nonfiction.

Fiction passages can include short stories, poems, novels, or even personal essays. Nonfiction passages can include informative or persuasive essays and can cover topics ranging from the humanities to the sciences.

The Reading section features questions that highlight the four major topics below. The questions on the test will **not** indicate which topic is being tested. Five answer choices are presented with letters A through E. The questions are **NOT** ordered according to level of difficulty.

- *Main Idea/Theme* – What is the general message, premise, or idea of the passage? What is the author trying to tell the reader?
- *Details* – What happens in the passage, and why? What do certain words or phrases mean?
- *Inferences* – What are some conclusions that can be drawn from the passage? What can a reader infer based on the passage?
- *Mood, Tone, Style, and Figurative Language* – What feeling is the author trying to convey? How does the author use figurative language, such as metaphors and similes, to establish a mood, or set a tone?

In This Practice Book

There are 20 passages in this section from a variety of sources, which reflect what students will see on the actual Middle Level SSAT. These passages have been divided into Fiction and Nonfiction sections. The corresponding questions will test students' ability to pinpoint some or all of the major topics that were outlined above.

We recommend that students practice several passages per week in preparing for the exam.

See an unfamiliar word? Look it up! Many words and concepts in this section might be challenging. It's a good habit to keep a list of vocabulary learned in passages, questions, or answer choices. Write down the definition of the word and use it in a sentence. Other notes that might help are positive or negative associations of the word, root words, or any phrases that are commonly encountered with the new word.

Tutorverse Tips!

Practice Active Reading

We recommend that you read the passage first before attempting to answer the questions. As you read, underline or circle key information like main ideas. Draw arrows between related ideas, or examples that support main ideas. Consider outlining the passage to get a sense for the structure of the passage as well as how the different parts of the passage are related to each other.

Because the questions on each passage will be similar to those that you have practiced, you can keep an eye out for important themes and ideas as you read. This will help save time when you answer the questions.

Identifying Main Ideas & Themes

Think about what the main idea might be as you read the passage. Ask yourself these questions as you read:

- What is the point of this passage? What is the author trying to tell me?
- What is the author's point of view on the topic?
- Is there a lesson or moral that I am supposed to learn from the passage?

Referring Back to the Text

- When the question refers back to the text with a quotation, make sure to read a little bit before and after the quoted text. Many times, the quote itself can have an ambiguous meaning if read by itself. Therefore, use context clues to help answer the questions.
- The same advice applies to questions that ask about a word or phrase's meaning. Use context clues, as the word or phrase will almost certainly have more than one possible definition.

Prove It!

Think you have the right answer? Prove it! You should be able to cite evidence from the text to support your answer for <u>every single question</u> – even inference questions! If you can't prove it to yourself, you probably haven't picked the right choice. Ask yourself, "How do I know this is true? What evidence is there from the passage that I can point to?"

Guessing

Knowing whether or not to guess can be tricky. The Middle Level SSAT gives you 1 point for each question that is answered correctly. However, if you answer a question incorrectly, ¼ of a point will be deducted from your total score (see example below).

The formula for determining the raw score is:
(Number of Questions Answered Correctly × 1) – (Number of Questions Answered Incorrectly × ¼)

No points will be awarded or deducted for questions that are left unanswered. Therefore, answer easy questions first, and come back to tougher questions later.

<u>How Guessing Impacts A Score</u>

Since there are 150 scored questions (167 total questions – 1 writing prompt – 16 experimental questions), the highest possible raw score is 150 points. This would be awarded to students who answer all 150 questions correctly.

If a student answers 110 questions correctly **but answers 40 questions incorrectly**, he or she will have earned (110 × 1) – (40 × ¼), or 110 – 10 = 100 points.

If a student answers 110 questions correctly **but leaves 40 questions unanswered**, he or she will have earned (110 × 1) – (40 × 0), or 110 – 0 = 110 points.

Therefore, it is important to refrain from making wild guesses. Instead, try to use process of elimination to make an educated guess.

Fiction

This section contains fiction passages. These passages have been adapted from short stories, novels, and poems, to help students become comfortable with the types of passages they will encounter on the actual test. Carefully read each passage and then answer the questions about it. For each question, select the choice that best answers the question based on the passage.

Passage #1

    Suddenly he became tense. Sound, sight, and smell had made him alert. His hand went back to touch the old man and the pair stood still. There arose a crackling sound up ahead, at one side of the top of the hill. The boy's gaze became fixed on the tops of the agitated bushes. Then, a large bear – a grizzly – crashed into view. It likewise stopped
5 abruptly at the sight of the humans. He did not like them, and growled irritably. Slowly the boy fitted his arrow to the bow, and slowly he pulled the bowstring taut. But he never removed his eyes from the bear.
    The old man stood as quietly as the boy and peered at the danger from under the brim of his hat. For a few seconds this mutual scrutinizing went on. Finally, as the bear
10 betrayed a growing annoyance, the boy, with a movement of his head, indicated that the old man must step aside from the trail and go down the hill. The boy followed, going backward, still holding the bow taut and ready. They waited until a crashing among the bushes from the opposite side of the hill told them the bear had gone on. The boy grinned as they headed back to the trail.
15 "A big one, Grandpa," he chuckled.
    The old man shook his head.

1. Which of the following best states the main idea of the passage?
   (A) A conflict reaches a violent ending.
   (B) A character acts foolishly in the face of danger.
   (C) A dangerous situation is handled sensibly.
   (D) Two friends slowly grow to distrust each other.
   (E) One character leads another character into danger.

2. According to the passage, what happens immediately after the bear growls (line 5)?
   (A) The pair make a joke and laugh.
   (B) The boy resorts to violence.
   (C) The companions turn around and run away.
   (D) The boy takes a precaution but remains calm.
   (E) The old man tells the boy about what to do next.

3. As used in the sentence, "betrayed" (line 10) most nearly means
   (A) deceived
   (B) concealed
   (C) revealed
   (D) cheated
   (E) released

4. The mood of the story is best described as which of the following?
   (A) comedic
   (B) uncertain
   (C) scholarly
   (D) suspenseful
   (E) overjoyed

5. It can be inferred from the passage that the boy
   (A) refuses to back down from danger.
   (B) is far more frightened than the old man.
   (C) has never encountered a grizzly bear before.
   (D) is quick to make irresponsible decisions.
   (E) has a sense of humor about the conflict with the bear.

Passage #2

Miss Archer called Lucy to her side and said, "Rosamund seems like such a nice girl!"

"Oh! don't begin by praising her," said Lucy. "I don't think I can stand it."

"What's the matter, my dear? Aren't you Lucy Merriman, daughter of Mrs.
5 Merriman and the Professor?"

"I am."

"And hasn't this house always been your home?"

"I was born here," said Lucy almost in tears.

"Then, of course you would feel strange at first with all these girls scattered about
10 the place. But when classes begin, you'll feel differently. You'll have to take your place with the others in classes, since everything will be conducted like a real school."

"I'll do anything you want," said Lucy, turning white with despair. "I'll do anything in all the world if you will promise me one thing."

Miss Archer desired to say, "Why should I promise you anything?" but she knew
15 human nature and guessed that Lucy was feeling troubled.

"Tell me what you wish," she said.

"I don't want you to make a favorite of Rosamund. She's begun to upset everything already. Last night, she started with all the living room arrangements and then her bedroom furniture afterwards. Today, the other girls have done nothing but obey her.
20 If this goes on, how can we maintain order?"

Miss Archer looked thoughtful.

1. It can be inferred from the passage that Lucy feels which of the following towards Rosamund?
   (A) sympathy
   (B) kindness
   (C) understanding
   (D) indifference
   (E) dislike

2. The passage primarily focuses on
   (A) the class schedule at a new school
   (B) an exchange between a girl and a teacher
   (C) two characters making a vow to each other
   (D) the positive qualities of a new student
   (E) a bitter conflict with no clear resolution

3. According to Miss Archer, Lucy is most likely upset because
   (A) she is in conflict with the Professor
   (B) she does not like school
   (C) she feels betrayed by Rosamund
   (D) her childhood home is now hosting other girls
   (E) she has been unable to complete her lessons

4. Which of the following best describes Rosamund?
   (A) She is well behaved.
   (B) She does not like Miss Archer.
   (C) She is jealous of Lucy's status.
   (D) She refuses to obey instructions.
   (E) She has earned the respect of the other girls.

5. It can be inferred from the passage that Miss Archer
   (A) is suspicious of Rosamund
   (B) is not familiar with Lucy's family life
   (C) does not care about Lucy's feelings
   (D) is willing to obey any of Lucy's wishes
   (E) has a strong understanding of human emotion

Passage #3

      I was now beginning to grow handsome; my coat had grown fine and soft, and it was a beautiful, bright black. I had one white foot and a pretty white star on my forehead. Everyone thought I was quite handsome, but my master would not sell me until I was four years old. He said boys should not work like men, and so colts should not work like
5 horses until they are grown up.
      When I was finally four years old, Squire Gordon came to inspect me. He examined my eyes, my mouth, and my legs, and then made me walk, trot, and gallop in front of him. He seemed to like me, and said, "When he has been well broken in, he will do very well." My master said he would break me in himself, as he did not want me to
10 become frightened or hurt. He did not delay, and the next day he began with my training.
      Not everyone may know what breaking in is; therefore, I will describe it. It means to teach a horse to wear a saddle and bridle, and to carry on his back a man, woman or child. More importantly, the horse must move exactly how the rider wishes, and move quietly. Besides this, he has to learn to wear a collar, a crupper, and a breeching, and to
15 stand still while they are put on. He must learn to have a cart or a carriage fixed behind, and he must be able to go fast or slow, just as his driver wishes. He must never become alarmed at what he sees, nor speak to other horses, nor bite, nor kick, nor have any will of his own. He must always do his master's will, even though he may be very tired or hungry. The worst of all is, when his harness is once on, he may neither jump for joy nor
20 lie down for weariness.

1. The passage is primarily about
   (A) how to break in a horse.
   (B) the most desirable qualities of a horse.
   (C) Squire Gordon's interaction with a horse.
   (D) a significant moment in a horse's life.
   (E) the relationship between a horse and its master.

2. According to the passage, a broken in horse should be able to do all of the following EXCEPT:
   (A) wear a breeching
   (B) not become distracted
   (C) obey its rider's demands
   (D) pull something behind it
   (E) gallop without becoming tired

3. The "I" referred to throughout the passage is
   (A) a horse.
   (B) the author.
   (C) Squire Gordon.
   (D) a horse's master.
   (E) a handsome man.

4. In line 2, the phrase "beautiful, bright black" is an example of
   (A) alliteration
   (B) hyperbole
   (C) allusion
   (D) metaphor
   (E) onomatopoeia

5. According to the passage, one limitation of being broken in is that the horse
   (A) is always tired.
   (B) loses its freedom.
   (C) is no longer unique.
   (D) must leave his master.
   (E) becomes less valuable to others.

Passage #4

Matthew enjoyed the ride except when he met women and had to nod to them, for on Prince Edward Island, you were supposed to nod to everyone you met on the road, whether you know them or not.

Matthew dreaded all women except Marilla and Mrs. Rachel. He had an
5 uncomfortable feeling that all other women were secretly laughing at him. He may have been right in thinking so, for he was an odd-looking person. He had an ungainly figure and long, iron-gray hair that touched his stooping shoulders, and a full, soft brown beard which he had since he was twenty.

When he reached Bright River there was no sign of any train. Matthew thought he
10 was too early, so he tied down his horse and walked to the station house. The long platform was almost deserted except for a girl who was sitting on a bench at the extreme end. Matthew walked past her as quickly as possible without looking at her. Had he looked at her he would have certainly noticed that she was tense and full of expectation. It would be clear to any passerby that she was sitting there waiting for something or
15 somebody.

Matthew encountered the stationmaster locking up the ticket office to go home for supper, and asked him if the five-thirty train would soon be arriving.

"The five-thirty train has been in and gone half an hour ago," answered the brisk official. "But there was a passenger dropped off for you – a little girl." The official shook
20 his head. "She's sitting out there on the bench. I asked her to go into the ladies' waiting room, but she informed me that she preferred to stay outside. I can't imagine why!"

1. As used in line 6, the word "ungainly" most nearly means
   (A) thin
   (B) bright
   (C) elegant
   (D) standard
   (E) awkward

2. According to the passage, Matthew fears that women on Prince Edward Island
   (A) remember him as a boy
   (B) assume he has a bad temper
   (C) secretly wish to speak to him
   (D) are teasing him behind his back
   (E) do not understand why he is not married

3. In the second paragraph, the writer describes
   (A) Mrs. Rachel's background
   (B) Matthew's peculiar appearance
   (C) the location of the Bright River station
   (D) the person Matthew is expecting to meet
   (E) how residents of Prince Edward Island are expected to behave

4. It can be inferred from the passage that Matthew walks past the waiting girl because he
   (A) is embarrassed for being late
   (B) is worried that she knows him
   (C) is uncomfortable around women
   (D) must first speak with the station official
   (E) never interacts with anyone but Marilla and Mrs. Rachel

5. The author implies which of the following about the little girl?
   (A) She is Mrs. Rachel's daughter.
   (B) She is eager to meet Matthew.
   (C) She is late to her next appointment.
   (D) She does not have a ticket so cannot wait inside.
   (E) She is nervous about who she is supposed to meet.

Passage #5

> Under a wall of bronze,
> Where beeches dip and trail
> Their branches in the water;
> With red-tipped head and wings—
> 5 A beaked ship under sail—
> There glides a single swan.
>
> Under the autumn trees
> He goes. The branches quiver,
> Dance in the wraith-like water,
> 10 Which ripples beneath the sedge
> With the slackening furrow that glides
> In his wake when he is gone:
> The beeches bow dark heads.
>
> Into the windless dusk,
> 15 Where in mist great towers stand
> Guarding a lonely strand,
> That is bodiless and dim,
> He speeds with easy stride;
> And I would go beside,
> 20 Till the low brown hills divide
> At last, for me and him.

1. Which statement best summarizes what happens in the poem?
   (A) A bird is seen swimming out of view.
   (B) A ghost is seen floating above still water.
   (C) A person walks alone looking at tall towers.
   (D) Trees are growing and dancing in the wind.
   (E) The water near a beach is seen moving beautifully.

2. The "wall of bronze" (line 1) most likely refers to the
   (A) color of the distant hills
   (B) color of a valuable metal
   (C) light from the setting sun
   (D) appearance of autumn leaves
   (E) mist that makes it hard to see

3. The "ship under sail" mentioned in line 5 refers to
   (A) a boat
   (B) an animal
   (C) the author
   (D) the appearance of trees
   (E) the movement of water

4. Which of the following is an example of a personification?
   (A) line 3
   (B) line 6
   (C) line 7
   (D) line 13
   (E) line 14

5. The mood of lines 14-17 is best described as
   (A) joyous
   (B) violent
   (C) mellow
   (D) brooding
   (E) embarrassed

Passage #6

> In that great journey of the stars through space
>   About the mighty, all-directing Sun,
>   The pallid, faithful Moon, has been the one
> Companion of the Earth. Her tender face,
> 5 Pale with the swift, keen purpose of that race,
>   Which at Time's natal hour was first begun,
>   Shines ever on her lover as they run
> And lights his orbit with her silvery smile.
>
>   Sometimes such passionate love doth in her rise,
> 10   Down from her beaten path she softly slips,
>   And with her mantle veils the Sun's bold eyes,
>   Then in the gloaming finds her lover's lips.
> While far and near the men our world call wise
>   See only that the Sun is in eclipse.

1. The mood of the poem is best described as
   (A) sad
   (B) angry
   (C) romantic
   (D) frustrated
   (E) disappointed

2. It can be inferred that the relationship between the Moon and the Earth in the poem is best described as
   (A) intimate
   (B) hostile
   (C) confused
   (D) erratic
   (E) maternal

3. Which of the following best describes the main idea of the first stanza of the poem?
   (A) The all-directing glow of the Sun acts as the faithful companion of the Moon.
   (B) The pallid moon lights up all journeys through space.
   (C) The Moon has been the Earth's most faithful companion since the beginning of time.
   (D) The mighty Sun has been orbiting the universe with a silvery smile.
   (E) Scientists and poets have different ways of interpreting a solar eclipse.

4. What is the meaning of the word "beaten" as used in line 10?
   (A) defeated
   (B) pummeled
   (C) tenderized
   (D) well-trodden
   (E) untouched

5. The figure of speech represented in line 8 is
   (A) metaphor
   (B) personification
   (C) hyperbole
   (D) alliteration
   (E) simile

Passage #7

> The day is fresh-washed and fair, and there is a smell of tulips and narcissus in the air.
> The sunshine pours in at the bath-room window and bores through the water in the bath-tub in lathes and planes of greenish-white. It cleaves the water into flaws like a
> 5 jewel, and cracks it to bright light.
> Little spots of sunshine lie on the surface of the water and dance, dance, and their reflections wobble deliciously over the ceiling; a stir of my finger sets them whirring, reeling. I move a foot and the planes of light in the water jar. I lie back and laugh, and let the green-white water, the sun-flawed beryl water, flow over me. The day is almost too
> 10 bright to bear, the green water covers me from the too bright day. I will lie here awhile and play with the water and the sun spots. The sky is blue and high. A crow flaps by the window, and there is a whiff of tulips and narcissus in the air.

1. The style of the poem is best described as
   (A) descriptive
   (B) persuasive
   (C) song-like
   (D) essay-like
   (E) historical

2. The word "beryl" (line 9) most likely describes a quality of
   (A) smell
   (B) sight
   (C) hearing
   (D) speech
   (E) imagination

3. Which of the following best summarizes what happens in the poem?
   (A) A person enjoys a relaxing bath.
   (B) Fresh tulips and narcissus smell nice.
   (C) Rain darkens an otherwise beautiful day.
   (D) A bird disturbs an otherwise relaxing experience.
   (E) A person buys jewels, which shine like a bright light.

4. According to the poem, the narrator laughs (line 8)
   (A) out of simple enjoyment
   (B) at a joke that she remembers
   (C) at the crow flapping its wings
   (D) because she hears something funny
   (E) at the strange sights she sees through the bathroom window

5. The feeling the poem evokes is one of
   (A) fear
   (B) surprise
   (C) pleasure
   (D) nervousness
   (E) depression

Passage #8

> The east is yellow as a daffodil.
> Three steeples—three stark swarthy arms—are thrust
> Up from the town. The gnarlèd poplars thrill
> Down the long street in some keen salty gust—
> 5   Straight from the sea and all the sailing ships—
> Turn white, black, white again, with noises sweet
> And swift. Back to the night the last star slips.
> High up the air is motionless, a sheet
> Of light. The east grows yellower apace,
> 10 And trembles: then, once more, and suddenly,
> The salt wind blows, and in that moment's space
> Flame roofs, and poplar-tops, and steeples three;
> From out the mist that wraps the river-ways,
> The little boats, like torches, start ablaze.

1. The speaker's tone is best described as
   (A) worried
   (B) critical
   (C) lively
   (D) negative
   (E) frightened

2. When the speaker states that "the little boats, like torches, start ablaze" (line 14), he is suggesting that the boats are
   (A) sinking
   (B) on fire
   (C) brightened by lightbulbs
   (D) being struck by lightning
   (E) reflecting in the sun

3. Which of the following best describes what happens in the poem?
   (A) Little boats burn near the shore.
   (B) Lightning causes trees to catch on fire.
   (C) The sun rises and brightens a coastal scene.
   (D) A powerful storm blows through a quiet town.
   (E) The narrator describes his love for sailing.

4. The figure of speech represented by line 1 is
   (A) simile
   (B) metaphor
   (C) personification
   (D) alliteration
   (E) irony

5. A good explanation for the sequence of events in lines 1-4 is
   (A) Flowers sway in the salty breeze near trees and churches.
   (B) The sun rises to the east, church rooftops point into the sky, and trees sway in a sea breeze.
   (C) People walk in the street of the town, picking yellow daffodils in the sun.
   (D) Trees line and daffodils grow around three churches in a town.
   (E) Salty gusts cool down a church congregation in the morning.

Passage #9

> We were very tired, we were very merry—
> We had gone back and forth all night on the ferry.
> It was bare and bright, and smelled like a stable—
> But we looked into a fire, we leaned across a table,
> 5   We lay on a hill-top underneath the moon;
> And the whistles kept blowing, and the dawn came soon.
>
> We were very tired, we were very merry—
> We had gone back and forth all night on the ferry;
> And you ate an apple, and I ate a pear,
> 10 From a dozen of each we had bought somewhere;
> And the sky went wan, and the wind came cold,
> And the sun rose dripping, a bucketful of gold.
>
> We were very tired, we were very merry,
> We had gone back and forth all night on the ferry.
> 15 We hailed, "Good morrow, mother!" to a shawl-covered head,
> And bought a morning paper, which neither of us read;
> And she wept, "God bless you!" for the apples and pears,
> And we gave her all our money but our subway fares.

1. Which words best capture the overall mood of the poem?
    (A) "We were very tired, we were very merry" (lines 1, 7, and 13)
    (B) "We had gone back and forth all night on the ferry" (lines 2, 8, and 14)
    (C) "It was bare and bright, and smelled like a stable" (line 3)
    (D) "And the sky went wan, and the wind became cold" (line 11)
    (E) "And brought a morning paper, which neither of us read" (line 16)

2. The first stanza (lines 1-6) suggests that the ferry was
    (A) cold yet comfortable
    (B) dark and dangerous
    (C) empty and foul-smelling
    (D) barren though clean
    (E) vibrant and full

3. The metaphor in line 12 refers to
    (A) dawn
    (B) nightfall
    (C) rainfall
    (D) a flower
    (E) treasure

4. Which of the following best describes the sequence of events in the third stanza (lines 13-18) of the poem?
    (A) A group sells newspapers to raise money for the ferry.
    (B) After riding the ferry, a group decided to travel on the subway all night.
    (C) After riding the ferry all night, a group buys fruit to snack upon.
    (D) Before giving away apples to a stranger, a group rides the subway.
    (E) A group greets another person and gives her money and food.

5. What is the main idea of the entire poem?
    (A) A sunrise puts an end to a night of fun and games.
    (B) A mother frets over her children, who rarely visit.
    (C) A group stays up all night, having fun and enjoying food.
    (D) A group of friends light a fire and spends time playing outside.
    (E) A group rides the subway all night, eating apples and pears.

Passage #10

> It winds along the face of a cliff
> This path which I long to explore,
> And over it dashes a waterfall,
> And the air is full of the roar
> 5 And the thunderous voice of waters which sweep
> In a silver torrent over some steep.
>
> It clears the path with a mighty bound
> And tumbles below and away,
> And the trees and the bushes which grow in the rocks
> 10 Are wet with its jeweled spray;
> The air is misty and heavy with sound,
> And small, wet wildflowers star the ground.
>
> Oh! The dampness is very good to smell,
> And the path is soft to tread,
> 15 And beyond the fall it winds up and on,
> While little streamlets thread
> Their own meandering way down the hill
> Each singing its own little song, until
>
> I forget that 't is only a pictured path,
> 20 And I hear the water and wind,
> And look through the mist, and strain my eyes
> To see what there is behind;
> For it must lead to a happy land,
> This little path by a waterfall spanned.

1. What is the overall main idea of the poem?
   (A) Rivers are a beautiful place to visit.
   (B) The waterfall is a powerful force of nature.
   (C) A hiker climbs a cliff to explore a beautiful place.
   (D) An artist demonstrates how to paint a beautiful picture.
   (E) The viewer of a picture describes a moving and powerful experience.

2. What is the best summary of the first stanza?
   (A) The speaker finds a dangerous place in nature.
   (B) The speaker describes a scene in detail.
   (C) The speaker compares the quality of one place with another.
   (D) The speaker gets caught in a thunderstorm while hiking.
   (E) The speaker discusses how best to cross a silver torrent.

3. The author's tone in the third stanza is best described as
   (A) miserable
   (B) serious
   (C) yearning
   (D) stressed
   (E) panicked

4. The "dampness" (line 13) is caused by a
   (A) lake
   (B) flood
   (C) waterfall
   (D) thunderstorm
   (E) melting glacier

5. The phrase "jeweled spray" (line 10) most likely refers to a
   (A) muddy river
   (B) crystal clear creek
   (C) flowing fountain
   (D) lightning-fast stream
   (E) rainbow-colored mist

Nonfiction

This section contains nonfiction passages that can explain an idea or attempt to persuade the reader. Carefully read each passage and then answer the questions about it. For each question, select the choice that best answers the question based on the passage.

Passage #1

The diseases vaccines prevent can be dangerous, and sometimes deadly. Vaccines reduce the risk of infection by working with the body's natural defenses to safely develop immunity to disease.

When germs, such as bacteria or viruses, invade the body, they attack and
5   multiply. This is known as an infection, and the infection is what causes illness. The immune system then has to fight the infection. Once it fights off the infection, the body has a supply of cells, or antibodies, that help recognize and fight that disease in the future.

Vaccines help develop immunity by imitating an infection, but this "imitation" infection does not cause illness. Instead, it causes the immune system to develop the same
10 response as it does to a real infection so the body can recognize and fight the vaccine-preventable disease in the future. Sometimes, after getting a vaccine, the imitation infection can cause minor symptoms, such as fever. Such minor symptoms are normal and should be expected as the body builds immunity.

Many vaccines are given to children, whose immune systems are not usually as
15 strong as those of adults. Even some adults, however, require vaccines. Over time, people require "booster" shots to help maintain immunity to certain diseases. In addition, people who travel to new places may encounter new diseases, and are wise to be vaccinated beforehand.

1. The best title for this passage would be
   (A) "Vaccines for Travelers"
   (B) "How Vaccines Actually Work"
   (C) "Beware of Vaccination: Dangerous Side Effects"
   (D) "Child Vaccination Techniques"
   (E) "Doctor's Recommendations: To Vaccinate or Not Vaccinate"

2. The word "immunity" (line 3) most likely means
   (A) prevention
   (B) illuminate
   (C) cooperation
   (D) resistance
   (E) revenge

3. According to the passage, what causes antibodies to develop?
   (A) the body fighting off an infection
   (B) being careful to never get sick
   (C) minor symptoms such as fever
   (D) the multiplication of bacteria and viruses
   (E) a child's natural aging process

4. Which of the following statements is NOT supported by the passage?
   (A) Children must get vaccines multiple times to maintain immunity.
   (B) Vaccines support the work of the natural immune system.
   (C) Side effects of vaccines include fever and other minor symptoms.
   (D) Vaccines imitate infections and cause illnesses.
   (E) Once vaccinated, antibodies will recognize and fight the disease in the future.

5. It can be inferred from the passage that the author believes that
   (A) vaccines are expensive but worth it
   (B) the pros of vaccines outweigh the cons
   (C) the cons of vaccines outweigh the pros
   (D) whether or not to vaccinate their child is the parent's choice
   (E) vaccines are highly dangerous or even deadly

Passage #2

>   Those who came before us made certain that this country rode the first waves of the industrial revolutions. They ensured that this country led the world in modern inventions. They made sure this country was a pioneer of nuclear power. And this generation does not intend to founder in the backwash of the coming age of space.
>
> 5   We intend to be the trailblazers of this new era, as the eyes of the world now look into space, to the moon and to the planets beyond. And we have vowed that we shall not see it governed by a hostile flag of conquest, but by the United States and its banner of freedom and peace. We shall not allow space to be filled with weapons of mass destruction, but with instruments of knowledge and understanding.
>
> 10 The vows of this Nation can only be fulfilled if we in this Nation are first, and, therefore, we intend to be first. In short, our leadership in science and industry, our hopes for peace and security, our obligations to ourselves as well as others, all require us to make this effort. We must solve these mysteries, and solve them for the good of all men. We must become the world's leading space-faring nation.
>
> 15 And so, we choose to go to the moon. We choose to go to the moon in this decade not because it is easy, but because it is hard. Because that goal will serve to organize and measure the best of our energies and skills. Because that challenge is one that we are willing to accept, one we are unwilling to postpone, and one which we intend to overcome.

1. We have vowed that we shall not see it governed by a hostile flag of conquest" (lines 6-7) reflects a tone that is
   (A) bitter
   (B) thankful
   (C) fearful
   (D) patriotic
   (E) aggressive

2. Which of the following is the best title for the selection?
   (A) "The First Waves of the Industrial Revolution"
   (B) "We Choose to Go to the Moon"
   (C) "No More Weapons of Mass Destruction"
   (D) "Leading Nation of the Free World"
   (E) "Do Not Postpone Freedom and Peace"

3. The word "founder" (line 4) most likely means
   (A) creator
   (B) sink
   (C) float
   (D) patriot
   (E) blossom

4. The author's charge to "make this effort" (line 13) is likely inspired by his claim that the United States
   (A) has stopped making efforts to invest in space exploration
   (B) needs more weapons of mass destruction
   (C) needs to ride the wave of the industrial revolution first
   (D) has an obligation to maintain peace in space as a leader in the world
   (E) is not ready for the coming age of space and must transform itself into a trailblazer

5. The author believes the Nation should "choose to go to the moon" (line 15) for all of the following reasons EXCEPT that it will
   (A) help solve scientific mysteries
   (B) organize and unite national energies
   (C) be easy and straightforward
   (D) measure various skills
   (E) present a challenge

Passage #3

Climate change has caused enormous damage. Even short-term changes – the volcanic eruptions near Crete 3,600 years ago, or the Tambora volcano eruptions in Indonesia, in 1815 – can have disastrous effect. And I'm talking about changes of one or two degrees. Today, we're setting forces in motion that can alter temperatures three or
5   four times as much.

I got a little taste of what the devastation could be like last spring during the floods that destroyed so many communities in the Midwest. It's one thing to read about these possibilities in scientific journals. It's another to fly over whole towns and see tops of houses and street signs poking above the water – or row a boat down what used to be
10 streets. Or talk to people who have lost literally every stick of furniture they own. And we can expect more and more floods just like this if we don't cut greenhouse gases. This isn't just the scientific view.

Take this quote: "Even a modest 0.9-degree Fahrenheit increase in average global temperature by the year 2010 could produce a 20-day extension of the hurricane season,
15 a 33% jump in hurricane landfalls in the U.S., an increase in the severity of the storms and a 30% annual rise in U.S. catastrophic losses from storms."

Written by a tree-hugging enviro? Nope. A study by the insurance giant, the Travelers Corporation, based in Hartford.

So it's not just alarmists who say that the climate change will be devastating and
20 we should try to prevent it. It's common sense. And it's good business sense, too. I challenge all of you here, facility managers, farmers, lawmakers and executives: we must come together to make it work.

1. This passage can best be described as
   (A) a persuasive speech
   (B) a presidential campaign
   (C) a scientific report
   (D) a private letter
   (E) a dramatic production

2. The word "alarmists" (line 19) most likely refer to those who
   (A) warn others wisely
   (B) panic unnecessarily
   (C) were once flood victims
   (D) love studying weather patterns
   (E) care more about money than nature

3. The main idea of the text is that
   (A) everyone must work together to solve a big problem
   (B) there are people who do not care about solving a big problem
   (C) insurance companies often disagree with environmentalists
   (D) most people would rather ignore a problem than solve it
   (E) solving big problems is the job of insurance companies

4. It can be inferred from the passage that the intended audience is most likely
   (A) students studying weather science
   (B) business owners, politicians and environmentalists
   (C) biology and chemistry scientists
   (D) people who do not believe in climate change
   (E) residents who live exclusively in the Midwest

5. The author uses the phrase "a tree-hugging enviro" (line 17) in order to
   (A) show disdain for environmentalists
   (B) refute a common argument against the speaker's point
   (C) make the previous quote less believable
   (D) show the speaker's grasp of slang
   (E) explain that climate change is not a serious threat

Passage #4

It may be difficult to fathom, but there was a long period in human history when fire was wild and uncontrolled. For millennia, fires caused by lightning would reduce forests to ash and clear vast grasslands. Though it often posed a significant threat to human life, these wildfires were not without their benefits. Fires large and small would
5   cook naturally occurring foods such as grasses and insects, making them easier to digest and more nutritious than when eaten raw.
   One of humanity's first attempts to control fire was to extend the life of these wildfires using animal dung. By using flammable materials such as this, early man could sustain and transfer fires to unburned areas and set fires of his own. In this way, our
10 ancestors could cook foods found in outlying areas or use the fire to create tools.
   One of the biggest turning points in human history occurred when mankind learned to set fires at will. The first man-made fires were sparked by the use of tools like the bow drill. When used properly, these prehistoric tools could generate intense friction, the heat from which was used to ignite moss, twigs, or other kindling. After carefully
15 excavating and studying these types of tools, scientists believe that mankind began making his own fires no earlier than 700,000 years ago and no later than 120,000 years ago.

1. The passage is primarily concerned with
   (A) exploring the diet of early man
   (B) describing humanity's first uses of fire
   (C) showing people found lightning to be useful
   (D) telling how the bow drill influenced human history
   (E) giving instructions on how to make a fire in the woods

2. According to the author, which of the following had the greatest impact on human history?
   (A) the use of the bow drill
   (B) employing prehistoric tools
   (C) cooking raw foods with fire
   (D) using animal dung to preserve fires
   (E) lightning storms clearing grasslands

3. The author uses "though" in line 3 to acknowledge that
   (A) wildfires were usually beneficial
   (B) lightning is always unpredictable
   (C) wildfires often destroyed valuable crops
   (D) humans preferred not to interact with fire
   (E) wildfires could often be dangerous to humans

4. The phrase "at will" (line 12) most nearly means
   (A) at random
   (B) as desired
   (C) in cooperation
   (D) without reason
   (E) with assistance

5. According to the passage, animal dung was useful as a
   (A) medicine
   (B) religious tool
   (C) food preservative
   (D) source of intense heat
   (E) tool for spreading fire

Passage #5

Many consider Thomas Edison to be one of the greatest inventors to have ever lived. After all, Edison is credited with developing such illustrious devices as the phonograph and the motion picture camera. Perhaps most famously, Edison is also recognized as the inventor of the electric light bulb, one of the most revolutionary
5   inventions of all time. However, few realize the truth; the light bulb was actually developed by an entire team of researchers!

In 1876, Edison established a research and development facility in Menlo Park, NJ. It was at this facility that Edison's employees carried out myriad tests that would eventually lead to the development of the light bulb. As they worked, Edison's
10   researchers relied upon the results of their many experiments as well as on the discoveries made by other inventors, like Nikola Tesla and Joseph Swan. When Edison's crowning achievement was finally finished, he filed a patent claiming discovery under his name alone.

Already considered to be a genius inventor by the public, Edison used this
15   celebrity to claim credit for the light bulb while simultaneously keeping others in the shadows.

1. The passage is primarily about
   (A) explaining a little-known fact
   (B) giving credit to specific people
   (C) the life and experiences of a famous inventor
   (D) describing the lasting impact of a creation
   (E) arguing that one invention is more important than another

2. The author would most likely agree with which of the following statements?
   (A) Edison should be given more credit for the invention of the light bulb.
   (B) Other scientists should be given credit for their work on the light bulb.
   (C) The phonograph, not the light bulb, is the most important of Edison's inventions.
   (D) People should no longer give Edison any credit for inventing the light bulb.
   (E) Most research facilities should be designed like Edison's Menlo Park establishment.

3. As used in line 8, "myriad" most nearly means
   (A) simple
   (B) random
   (C) numerous
   (D) confusing
   (E) mesmerizing

4. The "crowning achievement" mentioned in line 12 refers to
   (A) the light bulb
   (B) the phonograph
   (C) Tesla's discoveries
   (D) the motion picture camera
   (E) The research facility in Menlo Park

5. Which word could best replace "celebrity" as used in line 15 without changing the author's meaning?
   (A) actor
   (B) trust
   (C) fame
   (D) patent
   (E) invention

Passage #6

Many have heard of the *Mona Lisa*, Leonardo da Vinci's most famous work of art. Yet, few have stopped to consider why a simple painting of a silk merchant's wife receives millions of visitors each year. Some claim the painting's popularity owes to the subject's ambiguous smile, which appears to be either cheerful or suspicious, depending on whom
5   you ask. The truth, however, is far more dramatic.

On August 20, 1911, a man named Vincenzo Peruggia stole the *Mona Lisa* from the Louvre Museum in Paris. This brazen act took all of France by storm. Days turned into months, and there were no signs of the painting. Wild theories regarding the *Mona Lisa's* whereabouts flourished. News of the search for the missing painting plastered the front
10   page of newspapers for weeks. Even noted artist Pablo Picasso was investigated by police as a possible suspect. As public outcry over the missing painting increased, so too did the painting's popularity.

The thief was finally discovered and the painting recovered on December 11, 1913. By then, the *Mona Lisa* had captivated an entire nation. No fewer than 120,000
15   people went to see the painting in the first two days after it was returned to the Louvre.

1. As used in line 4, the word "ambiguous" most nearly means
   (A) stiff
   (B) calm
   (C) eager
   (D) gloomy
   (E) baffling

2. The author most likely references Pablo Picasso in order to
   (A) show the police as lazy
   (B) criticize Leonardo da Vinci's art
   (C) prove Peruggia's guilt as the thief
   (D) compare two noteworthy painters
   (E) illustrate the intensity of the search

3. The passage states that the *Mona Lisa* is popular because
   (A) of the subject's smile
   (B) of Vincenzo Peruggia's actions
   (C) many art critics love the painting
   (D) it was an example of a new style of art
   (E) of the Louvre Museum's frequent promotions

4. The tone of the passage can best be described as
   (A) argumentative
   (B) harsh
   (C) informative
   (D) skeptical
   (E) pessimistic

5. The author describes how newspapers (lines 9-10)
   (A) asked people to see the painting
   (B) spread rumors about Pablo Picasso
   (C) helped people learn about the painting
   (D) had a role in the painting's popularity
   (E) encouraged the police to catch the criminal

Passage #7

E-cigarettes, or electronic cigarettes, were first created in China in 2003. Though they were advertised as a healthier alternative to "Big Tobacco," this was just a marketing ploy. New research shows that these claims were dishonest: like traditional cigarettes, long term use of e-cigarettes has damaging side effects. Because of this, they should be
5   banned across the globe.
   Some brands of e-cigarettes can deliver deadly doses of nicotine, a toxic substance that can be poisonous in large amounts. In fact, the official nicotine content of some e-cigarette brands actually exceeds the average amount in traditional cigarettes, and studies have shown that e-cigarette labels may list less nicotine content than is actually
10 present. Additionally, many e-cigarettes contain a chemical known for causing "popcorn lung," a disease that results in inflammation of the lungs. Some scientists believe that like traditional cigarettes, e-cigarettes also negatively affect lung capacity, which puts users at a higher risk of experiencing a stroke or heart attack. There is clear evidence that e-cigarettes can be just as dangerous to our health as traditional cigarettes.
15 In total, there are 70 countries that are regulating e-cigarettes. For example, e-cigarettes are now illegal in the United Arab Emirates. They can be possessed, but not sold, in countries like Brazil. In nations like Honduras, they are banned from being smoked in public. Here in the United States, selling e-cigarettes to minors, as well as providing free samples, are both punishable offenses.
20 Unfortunately, these restrictions are not enough. If we do not create stronger laws to ban e-cigarette production, we may be contributing to yet another public health crisis.

1. The main argument of the passage is that e-cigarettes
   (A) are a healthier alternative to traditional cigarettes
   (B) should be banned entirely
   (C) should be legalized internationally
   (D) can cause strokes or heart attacks
   (E) are banned from being smoked in public in Honduras

2. The writer uses the phrase "Big Tobacco" (line 2) to refer to
   (A) the e-cigarette industry
   (B) popcorn lung
   (C) marketing ploys
   (D) the toxic effects of nicotine
   (E) traditional cigarettes

3. The tone of the claims made in this passage are best described as
   (A) depressing
   (B) humorous
   (C) indecisive
   (D) assertive
   (E) excited

4. The word "ploy" (line 3) most likely means
   (A) playful
   (B) tool
   (C) dishonest
   (D) damage
   (E) trick

5. The author uses all of the following claims to prove that "e-cigarettes can be just as dangerous to our health as traditional cigarettes" (lines 13-14) EXCEPT
   (A) "e-cigarettes can deliver deadly doses of nicotine" (line 6)
   (B) "the official nicotine content of some e-cigarette brands actually exceeds the average amount in traditional cigarettes" (lines 7-8)
   (C) "cigarettes contain a chemical known for causing popcorn lung" (line 10-11)
   (D) "e-cigarettes also negatively affect lung capacity" (line 12)
   (E) "there are 70 countries that are regulating e-cigarettes" (line 15)

Passage #8

Many people mix olive oil and vinegar to make salad dressing. If mixed vigorously, it may appear that the oil and the vinegar become one. However, the two liquids will eventually separate no matter how well mixed it may appear. This is because olive oil and vinegar are made of different types of molecules.

5   Vinegar is made up of acetic acid and water, both of which are polar molecules. Polarization results from an uneven distribution of electrons among atoms. A polar molecule has a slightly negative charge at one end and a slightly positive charge at the other end. Polar molecules attract other polar molecules because the negative charge of one molecule seeks out the positive charge of another molecule. Consequently, water and
10   acetic acid molecules join together, forming vinegar.

Oil, on the other hand, is a type of fat and is non-polar. Oils are made up of fatty acids that contain atoms with evenly distributed electrons. As a result, the atoms in oils do not have any charge, whether negative or positive. Like charged atoms, non-charged atoms like to stick together. This means that oil molecules repel vinegar molecules.

15   As a result, charged vinegar molecules race toward each other, while non-charged oil molecules strive to stay together. This is the reason why it's important to use oil and vinegar salad dressing while freshly mixed!

1. The author is primarily concerned with
   (A) how to make a tasty salad dressing
   (B) the types of oil and vinegar that mix well
   (C) describing why acetic acid is polarized
   (D) why charged molecules attract one another
   (E) the reason why olive oil and vinegar do not mix

2. It can be inferred from the passage that substances with evenly distributed electrons
   (A) do not group together
   (B) can be found in vinegar
   (C) are always called fatty acids
   (D) usually have either a positive or negative charge
   (E) are not attracted to polarized molecules

3. According to the passage, why are water and acetic acid drawn towards each other?
   (A) They both lack charges.
   (B) They both naturally occur in oils.
   (C) They both are polarized molecules.
   (D) They form an important ingredient in salad dressing.
   (E) They are both entirely made up of negatively charge molecules.

4. The author claims that olive oil and vinegar cannot be permanently mixed together because
   (A) neither can combine with water
   (B) one is polarized and the other non-polarized
   (C) polar molecules attract non-polar molecules
   (D) no person is strong enough to mix the two together
   (E) there aren't enough electrons for the two liquids to combine

5. The writer's style is best described as
   (A) flowery
   (B) slangy
   (C) educational
   (D) wordy
   (E) vague

Passage #9

The unmistakable *sssss* of a newly opened bottle of soda – this pleasing sound of refreshment seems unassuming enough but is actually caused by a complex series of chemical reactions.

At soda bottling factories, approximately 2.2 grams of carbon dioxide gas are
5 pumped into each bottle of flavored water. Once the bottle is sealed, the carbon dioxide, which wants to escape, has nowhere to go. This, in turn, creates a pressurized environment inside the bottle. Under pressure, the liquid in the bottle absorbs the carbon dioxide until it reaches a state of equilibrium. This is achieved when there is the same amount of gas dissolved in the liquid as there is in the space at the top of the bottle.
10 When a bottle of soda is opened – by a thirsty 6th grader, for example – this balance is broken. As the soda water is exposed to the normal pressure in the air, the pressure in the bottle drops, and the carbon dioxide escapes. As a result, bubbles of carbon dioxide form rapidly, rise to the surface, and burst, releasing the gas into the air.

1. The author's use of "*sssss*" in line 1 is an example of
   (A) metaphor
   (B) hyperbole
   (C) alliteration
   (D) onomatopoeia
   (E) personification

2. Which word could best replace "unassuming" as used in line 2 without changing the author's meaning?
   (A) certain
   (B) intricate
   (C) problematic
   (D) definite
   (E) simple

3. The passage was probably taken from a
   (A) literary journal
   (B) local newspaper
   (C) science textbook
   (D) magazine for chefs
   (E) catalogue of drinks

4. According to the passage, the gas in a bottle of soda achieves a state of equilibrium when
   (A) the liquid is poured into the bottle
   (B) water is combined with flavorings
   (C) when the bottle is opened for drinking
   (D) one can see gas bubbles rising through the liquid
   (E) there is the same amount of gas in the liquid as in the empty space in a bottle

5. It can be inferred from the passage that an unopened bottle of soda does not look bubbly because
   (A) the bottle is not pressurized
   (B) not all soda bottles are sealed properly
   (C) the soda has not yet reached equilibrium
   (D) some soda is not infused with carbon dioxide
   (E) carbon dioxide bubbles only form when equilibrium is upset

Passage #10

One of the biggest threats facing the environment today is plastic. Huge patches of plastic float on the surfaces of oceans around the world. One of them, the Great Pacific Garbage Patch, which floats between California and Hawaii, is almost three times the size of France! Plastic does not biodegrade easily. Most living things cannot eat and digest
5   plastic, so it does not break down the way that plants and animals do when they die. Instead, it remains in the environment for hundreds of years, where it can do serious harm to animals that eat it by mistake.

But scientists are learning that some creatures *can* eat plastic. In 2016, scientists in Japan discovered a new kind of bacteria living in a garbage dump. This species of
10 bacteria, *Ideonella sakaiensis*, breaks down a certain chemical found in soft drink bottles and uses it for food. Wax moth caterpillars have a similar ability. Beekeepers have long considered this species of caterpillar a pest because it feeds on the wax of their beehives. But when one beekeeper used a plastic bag to remove wax moth caterpillars from a beehive, she found that they had eaten holes in the bag!

15 Will these discoveries help us solve our plastic problem? It is not yet clear how to put these creatures to work for us. One wax moth caterpillar, for instance, can only eat a very tiny amount of plastic per day, and more than a million tons of plastic flow into the world's oceans each year. Yet scientists are hopeful. Researchers have found a way to enhance the chemical that *Ideonella sakaiensis* uses to digest plastic. It may be possible to
20 use this chemical or others like it to make plastic easier to recycle.

Nature is learning how to deal with plastic in the environment. Will we work with it, or against it?

1. According to the author, the worst thing about plastic for the environment is that it
   (A) collects in large amounts
   (B) injures wildlife that eat it
   (C) is very inexpensive to make
   (D) floats on the ocean's surface
   (E) is eaten by wax moth caterpillars

2. As used in the sentence, "biodegrade" (line 4) most nearly means
   (A) eat other living things
   (B) harm the environment
   (C) last for a very long time
   (D) be made by living things
   (E) be broken down by living things

3. Which of the following is the best title for the selection?
   (A) Saving the Ocean
   (B) The Beekeeper's Handbook
   (C) The Great Pacific Garbage Patch
   (D) A Natural Solution to Plastic Waste
   (E) The Lifecycle of *Ideonella sakaiensis*

4. Which sentences best describes how paragraph 2 supports the passage's main idea?
   (A) It describes two species of animal and what they eat.
   (B) It describes how scientists discover new animal species.
   (C) It shows that scientists still have a lot to learn about animals.
   (D) It describes new ways that plastic interacts with the environment.
   (E) It shows that cleaning up plastic will be easier than scientists expected.

5. It can be inferred from the passage that the author thinks that
   (A) people should use less plastic in their daily lives
   (B) wax moth caterpillars do more harm than good
   (C) only *Ideonella sakaiensis* can be used to clean up plastic
   (D) using wax moth caterpillars alone isn't enough to clean up plastic
   (E) there is too much plastic waste to bother trying cleaning it up

The Writing Sample

Overview

Before students begin the multiple-choice section of the Middle Level SSAT, they will be asked to complete a writing sample. This must be completed in 25 minutes. While this writing sample is **not** scored, it will be sent to the admissions officers at the schools to which students apply. The writing sample will be used by schools to assess a student's writing skills and learn more about him or her. A copy of the writing sample will **not** be included in the scores provided unless separately purchased from the SSAT.

On the Actual Test

Students will have a choice between two prompts: a creative piece, and an essay.

The creative prompt is intended to spark a student's imagination. It will ask the student to write a story based on a phrase. The essay prompt will ask the student to consider something academic, or to describe something about his or her life. Students should select the choice that speaks to him or her!

Students will have 25 minutes to write the sample. Consider mapping out time as follows:

3 minutes – *Plan*. Brainstorm ideas and jot down notes on the scrap paper that will be provided. Students should generally organize their samples into five paragraphs (introduction, body paragraphs, and conclusion).

20 minutes – *Write*. If students have planned well, the actual writing of the sample should be a breeze. Remember: big words and long sentences by themselves can't compensate for clearly and concisely communicated ideas!

2 minutes – *Proofread*. This is crucial! Reread the sample to look for and correct any punctuation, spelling, and grammar errors.

Tutorverse Tips!

Remember that the writing sample has two purposes. Schools want to see how well a student can write, but also want to learn something about the student as a person. Think of the writing sample as a written interview. If a student is asked to describe the best birthday party he or she has ever attended, the student may want to consider focusing on the experience itself and how it made him or her feel. Why was the birthday party special? How did the party make the student feel? Instead of focusing only on describing the events that happened, ask "So what?" – why were the events special?

Remember to plan thoroughly before writing and to proofread carefully when finished. The planning is important because admissions directors can identify well organized writing samples versus those samples that lack structure. Proofreading is important in order to remove careless mistakes – such as simple punctuation or spelling errors – that will reflect poorly on a student's writing skills.

How to Use This Section

Below are 10 essay prompts that have been grouped into 5 pairs. Treat each pair as a prompt you might see on the real test.

1. Set a timer for 25 minutes.
2. Choose one topic from the first pair. Think about which prompt allows you to best express yourself.
3. Make your notes on a separate piece of paper.
4. Write your essay on a separate sheet of lined paper.
5. Remember to proofread!

Schools would like to get to know you through an essay or story that you write. Choose one of the topics below that you find most interesting. Fill in the circle next to the topic of your choice. Then, write a story or essay based on the topic you chose.

Ⓐ Tell about a time you failed at something. What did you learn?

Ⓑ She thought she could get away with it.

Ⓐ Tell about the best teacher you've ever had. What made him or her so great?

Ⓑ He looked everywhere.

Ⓐ If you could wake up with a new skill, what would it be?

Ⓑ I approached the stage, my heart pounding.

Ⓐ Tell about a time you struggled and succeeded.

Ⓑ I was so happy, I could barely contain it.

Ⓐ If you became President of the United States, what is the first thing you would do?

Ⓑ I ran as fast as I could, but he was gaining on me.

Final Practice Test (Form B)

Overview

The first step to an effective study plan is to determine a student's strengths and areas for improvement. This first practice test assesses a student's existing knowledge and grasp of concepts that may be seen on the actual exam.

Keep in mind that this practice test will be scored differently from the actual exam. On the actual Middle Level SSAT, **certain questions will not count towards a student's actual score (i.e. the experimental section)**. Also, the student's score will be determined by comparing his or her performance with those of other students in the same grade. On this practice test, however, every question is scored in order to accurately gauge the student's current ability level. Therefore, **this practice test should NOT be used as a gauge of how a student will score on the actual test**. This test should only be used to help students develop a study plan, and may be treated as a diagnostic test.

Format

The format of this diagnostic practice test is similar to that of the actual exam and includes 16 questions in a mock-experimental section. **For practice purposes only, students should treat the mock experimental section of the diagnostic practice test as any other.**

The format of the diagnostic practice test is below.

Scoring	Section	Number of Questions	Time Limit
Unscored Section (sent to schools)	Writing Sample	1	25 minutes
Scored Section	5-Minute Break		
	Section 1: Quantitative	25	30 minutes
	Section 2: Reading	40	40 minutes
	10-Minute Break		
	Section 3: Verbal	60	30 minutes
	Section 4: Quantitative	25	30 minutes
	Total Scored Exam (Sections 1-4)	**150**	**2 hours, 10 minutes**
Unscored Section	Section 5: Experimental	16	15 minutes

Answering

Use the answer sheet provided on the next page to record answers. Students may wish to tear it out of the workbook.

Final Practice Test (Form B)

Section 1: Quantitative

1. Ⓐ Ⓑ Ⓒ Ⓓ Ⓔ
2. Ⓐ Ⓑ Ⓒ Ⓓ Ⓔ
3. Ⓐ Ⓑ Ⓒ Ⓓ Ⓔ
4. Ⓐ Ⓑ Ⓒ Ⓓ Ⓔ
5. Ⓐ Ⓑ Ⓒ Ⓓ Ⓔ
6. Ⓐ Ⓑ Ⓒ Ⓓ Ⓔ
7. Ⓐ Ⓑ Ⓒ Ⓓ Ⓔ
8. Ⓐ Ⓑ Ⓒ Ⓓ Ⓔ
9. Ⓐ Ⓑ Ⓒ Ⓓ Ⓔ
10. Ⓐ Ⓑ Ⓒ Ⓓ Ⓔ
11. Ⓐ Ⓑ Ⓒ Ⓓ Ⓔ
12. Ⓐ Ⓑ Ⓒ Ⓓ Ⓔ
13. Ⓐ Ⓑ Ⓒ Ⓓ Ⓔ
14. Ⓐ Ⓑ Ⓒ Ⓓ Ⓔ
15. Ⓐ Ⓑ Ⓒ Ⓓ Ⓔ
16. Ⓐ Ⓑ Ⓒ Ⓓ Ⓔ
17. Ⓐ Ⓑ Ⓒ Ⓓ Ⓔ
18. Ⓐ Ⓑ Ⓒ Ⓓ Ⓔ
19. Ⓐ Ⓑ Ⓒ Ⓓ Ⓔ
20. Ⓐ Ⓑ Ⓒ Ⓓ Ⓔ
21. Ⓐ Ⓑ Ⓒ Ⓓ Ⓔ
22. Ⓐ Ⓑ Ⓒ Ⓓ Ⓔ
23. Ⓐ Ⓑ Ⓒ Ⓓ Ⓔ
24. Ⓐ Ⓑ Ⓒ Ⓓ Ⓔ
25. Ⓐ Ⓑ Ⓒ Ⓓ Ⓔ

Section 2: Reading

1. Ⓐ Ⓑ Ⓒ Ⓓ Ⓔ
2. Ⓐ Ⓑ Ⓒ Ⓓ Ⓔ
3. Ⓐ Ⓑ Ⓒ Ⓓ Ⓔ
4. Ⓐ Ⓑ Ⓒ Ⓓ Ⓔ
5. Ⓐ Ⓑ Ⓒ Ⓓ Ⓔ
6. Ⓐ Ⓑ Ⓒ Ⓓ Ⓔ
7. Ⓐ Ⓑ Ⓒ Ⓓ Ⓔ
8. Ⓐ Ⓑ Ⓒ Ⓓ Ⓔ
9. Ⓐ Ⓑ Ⓒ Ⓓ Ⓔ
10. Ⓐ Ⓑ Ⓒ Ⓓ Ⓔ
11. Ⓐ Ⓑ Ⓒ Ⓓ Ⓔ
12. Ⓐ Ⓑ Ⓒ Ⓓ Ⓔ
13. Ⓐ Ⓑ Ⓒ Ⓓ Ⓔ
14. Ⓐ Ⓑ Ⓒ Ⓓ Ⓔ
15. Ⓐ Ⓑ Ⓒ Ⓓ Ⓔ
16. Ⓐ Ⓑ Ⓒ Ⓓ Ⓔ
17. Ⓐ Ⓑ Ⓒ Ⓓ Ⓔ
18. Ⓐ Ⓑ Ⓒ Ⓓ Ⓔ
19. Ⓐ Ⓑ Ⓒ Ⓓ Ⓔ
20. Ⓐ Ⓑ Ⓒ Ⓓ Ⓔ
21. Ⓐ Ⓑ Ⓒ Ⓓ Ⓔ
22. Ⓐ Ⓑ Ⓒ Ⓓ Ⓔ
23. Ⓐ Ⓑ Ⓒ Ⓓ Ⓔ
24. Ⓐ Ⓑ Ⓒ Ⓓ Ⓔ
25. Ⓐ Ⓑ Ⓒ Ⓓ Ⓔ
26. Ⓐ Ⓑ Ⓒ Ⓓ Ⓔ
27. Ⓐ Ⓑ Ⓒ Ⓓ Ⓔ
28. Ⓐ Ⓑ Ⓒ Ⓓ Ⓔ
29. Ⓐ Ⓑ Ⓒ Ⓓ Ⓔ
30. Ⓐ Ⓑ Ⓒ Ⓓ Ⓔ
31. Ⓐ Ⓑ Ⓒ Ⓓ Ⓔ
32. Ⓐ Ⓑ Ⓒ Ⓓ Ⓔ
33. Ⓐ Ⓑ Ⓒ Ⓓ Ⓔ
34. Ⓐ Ⓑ Ⓒ Ⓓ Ⓔ
35. Ⓐ Ⓑ Ⓒ Ⓓ Ⓔ
36. Ⓐ Ⓑ Ⓒ Ⓓ Ⓔ
37. Ⓐ Ⓑ Ⓒ Ⓓ Ⓔ
38. Ⓐ Ⓑ Ⓒ Ⓓ Ⓔ
39. Ⓐ Ⓑ Ⓒ Ⓓ Ⓔ
40. Ⓐ Ⓑ Ⓒ Ⓓ Ⓔ

Section 3: Verbal

1. Ⓐ Ⓑ Ⓒ Ⓓ Ⓔ
2. Ⓐ Ⓑ Ⓒ Ⓓ Ⓔ
3. Ⓐ Ⓑ Ⓒ Ⓓ Ⓔ
4. Ⓐ Ⓑ Ⓒ Ⓓ Ⓔ
5. Ⓐ Ⓑ Ⓒ Ⓓ Ⓔ
6. Ⓐ Ⓑ Ⓒ Ⓓ Ⓔ
7. Ⓐ Ⓑ Ⓒ Ⓓ Ⓔ
8. Ⓐ Ⓑ Ⓒ Ⓓ Ⓔ
9. Ⓐ Ⓑ Ⓒ Ⓓ Ⓔ
10. Ⓐ Ⓑ Ⓒ Ⓓ Ⓔ
11. Ⓐ Ⓑ Ⓒ Ⓓ Ⓔ
12. Ⓐ Ⓑ Ⓒ Ⓓ Ⓔ
13. Ⓐ Ⓑ Ⓒ Ⓓ Ⓔ
14. Ⓐ Ⓑ Ⓒ Ⓓ Ⓔ
15. Ⓐ Ⓑ Ⓒ Ⓓ Ⓔ
16. Ⓐ Ⓑ Ⓒ Ⓓ Ⓔ
17. Ⓐ Ⓑ Ⓒ Ⓓ Ⓔ
18. Ⓐ Ⓑ Ⓒ Ⓓ Ⓔ
19. Ⓐ Ⓑ Ⓒ Ⓓ Ⓔ
20. Ⓐ Ⓑ Ⓒ Ⓓ Ⓔ
21. Ⓐ Ⓑ Ⓒ Ⓓ Ⓔ
22. Ⓐ Ⓑ Ⓒ Ⓓ Ⓔ
23. Ⓐ Ⓑ Ⓒ Ⓓ Ⓔ
24. Ⓐ Ⓑ Ⓒ Ⓓ Ⓔ
25. Ⓐ Ⓑ Ⓒ Ⓓ Ⓔ
26. Ⓐ Ⓑ Ⓒ Ⓓ Ⓔ
27. Ⓐ Ⓑ Ⓒ Ⓓ Ⓔ
28. Ⓐ Ⓑ Ⓒ Ⓓ Ⓔ
29. Ⓐ Ⓑ Ⓒ Ⓓ Ⓔ
30. Ⓐ Ⓑ Ⓒ Ⓓ Ⓔ
31. Ⓐ Ⓑ Ⓒ Ⓓ Ⓔ
32. Ⓐ Ⓑ Ⓒ Ⓓ Ⓔ
33. Ⓐ Ⓑ Ⓒ Ⓓ Ⓔ
34. Ⓐ Ⓑ Ⓒ Ⓓ Ⓔ
35. Ⓐ Ⓑ Ⓒ Ⓓ Ⓔ
36. Ⓐ Ⓑ Ⓒ Ⓓ Ⓔ
37. Ⓐ Ⓑ Ⓒ Ⓓ Ⓔ
38. Ⓐ Ⓑ Ⓒ Ⓓ Ⓔ
39. Ⓐ Ⓑ Ⓒ Ⓓ Ⓔ
40. Ⓐ Ⓑ Ⓒ Ⓓ Ⓔ
41. Ⓐ Ⓑ Ⓒ Ⓓ Ⓔ
42. Ⓐ Ⓑ Ⓒ Ⓓ Ⓔ
43. Ⓐ Ⓑ Ⓒ Ⓓ Ⓔ
44. Ⓐ Ⓑ Ⓒ Ⓓ Ⓔ
45. Ⓐ Ⓑ Ⓒ Ⓓ Ⓔ
46. Ⓐ Ⓑ Ⓒ Ⓓ Ⓔ
47. Ⓐ Ⓑ Ⓒ Ⓓ Ⓔ
48. Ⓐ Ⓑ Ⓒ Ⓓ Ⓔ
49. Ⓐ Ⓑ Ⓒ Ⓓ Ⓔ
50. Ⓐ Ⓑ Ⓒ Ⓓ Ⓔ
51. Ⓐ Ⓑ Ⓒ Ⓓ Ⓔ
52. Ⓐ Ⓑ Ⓒ Ⓓ Ⓔ
53. Ⓐ Ⓑ Ⓒ Ⓓ Ⓔ
54. Ⓐ Ⓑ Ⓒ Ⓓ Ⓔ
55. Ⓐ Ⓑ Ⓒ Ⓓ Ⓔ
56. Ⓐ Ⓑ Ⓒ Ⓓ Ⓔ
57. Ⓐ Ⓑ Ⓒ Ⓓ Ⓔ
58. Ⓐ Ⓑ Ⓒ Ⓓ Ⓔ
59. Ⓐ Ⓑ Ⓒ Ⓓ Ⓔ
60. Ⓐ Ⓑ Ⓒ Ⓓ Ⓔ

Section 4: Quantitative

1. Ⓐ Ⓑ Ⓒ Ⓓ Ⓔ
2. Ⓐ Ⓑ Ⓒ Ⓓ Ⓔ
3. Ⓐ Ⓑ Ⓒ Ⓓ Ⓔ
4. Ⓐ Ⓑ Ⓒ Ⓓ Ⓔ
5. Ⓐ Ⓑ Ⓒ Ⓓ Ⓔ
6. Ⓐ Ⓑ Ⓒ Ⓓ Ⓔ
7. Ⓐ Ⓑ Ⓒ Ⓓ Ⓔ
8. Ⓐ Ⓑ Ⓒ Ⓓ Ⓔ
9. Ⓐ Ⓑ Ⓒ Ⓓ Ⓔ
10. Ⓐ Ⓑ Ⓒ Ⓓ Ⓔ
11. Ⓐ Ⓑ Ⓒ Ⓓ Ⓔ
12. Ⓐ Ⓑ Ⓒ Ⓓ Ⓔ
13. Ⓐ Ⓑ Ⓒ Ⓓ Ⓔ
14. Ⓐ Ⓑ Ⓒ Ⓓ Ⓔ
15. Ⓐ Ⓑ Ⓒ Ⓓ Ⓔ
16. Ⓐ Ⓑ Ⓒ Ⓓ Ⓔ
17. Ⓐ Ⓑ Ⓒ Ⓓ Ⓔ
18. Ⓐ Ⓑ Ⓒ Ⓓ Ⓔ
19. Ⓐ Ⓑ Ⓒ Ⓓ Ⓔ
20. Ⓐ Ⓑ Ⓒ Ⓓ Ⓔ
21. Ⓐ Ⓑ Ⓒ Ⓓ Ⓔ
22. Ⓐ Ⓑ Ⓒ Ⓓ Ⓔ
23. Ⓐ Ⓑ Ⓒ Ⓓ Ⓔ
24. Ⓐ Ⓑ Ⓒ Ⓓ Ⓔ
25. Ⓐ Ⓑ Ⓒ Ⓓ Ⓔ

Section 5: Experimental

1. Ⓐ Ⓑ Ⓒ Ⓓ Ⓔ
2. Ⓐ Ⓑ Ⓒ Ⓓ Ⓔ
3. Ⓐ Ⓑ Ⓒ Ⓓ Ⓔ
4. Ⓐ Ⓑ Ⓒ Ⓓ Ⓔ
5. Ⓐ Ⓑ Ⓒ Ⓓ Ⓔ
6. Ⓐ Ⓑ Ⓒ Ⓓ Ⓔ
7. Ⓐ Ⓑ Ⓒ Ⓓ Ⓔ
8. Ⓐ Ⓑ Ⓒ Ⓓ Ⓔ
9. Ⓐ Ⓑ Ⓒ Ⓓ Ⓔ
10. Ⓐ Ⓑ Ⓒ Ⓓ Ⓔ
11. Ⓐ Ⓑ Ⓒ Ⓓ Ⓔ
12. Ⓐ Ⓑ Ⓒ Ⓓ Ⓔ
13. Ⓐ Ⓑ Ⓒ Ⓓ Ⓔ
14. Ⓐ Ⓑ Ⓒ Ⓓ Ⓔ
15. Ⓐ Ⓑ Ⓒ Ⓓ Ⓔ
16. Ⓐ Ⓑ Ⓒ Ⓓ Ⓔ

The Tutorverse
www.thetutorverse.com

Writing Sample

Schools would like to get to know you through an essay or story that you write. Choose one of the topics below that you find most interesting. Fill in the circle next to the topic of your choice. Then, write a story or essay based on the topic you chose.

> Ⓐ If you could change one thing about your school, what would you change, and why?
>
> Ⓑ I could practically taste it.

Use this page and the next page to complete your writing sample.

SECTION 1
25 Questions

There are five suggested answers after each problem in this section. Solve each problem in your head or in the space provided to the right of the problem. Then, look at the suggested answers and pick the best one.

Note: Any figures or shapes that accompany problems in Section 1 are drawn as accurately as possible EXCEPT when it is stated that the figure is NOT drawn to scale.

Sample Question:

11 × 14 =	●ⒷⒸⒹⒺ
(A) 154	
(B) 196	
(C) 1,114	
(D) 1,554	
(E) 1,969	

DO WORK IN THIS SPACE

1. What is the value of the expression $40 - (8 \times 6 \div 12) \div 2^2$?
   (A) 39
   (B) 36
   (C) 24
   (D) 9
   (E) 1

2. Danica did not study for her first science test and scored an 80. She studied for her second science test and scored a 90. By what percent did her test score increase from her first science test to her second?
   (A) 9%
   (B) 10%
   (C) $11\frac{1}{9}$%
   (D) $12\frac{1}{2}$%
   (E) 25%

3. Compute the sum 12.4 + 1.247.
   (A) 13.647
   (B) 13.687
   (C) 13.87
   (D) 24.67
   (E) 24.87

4. A sewing machine can sew 120 stitches in one minute. At this rate, how many stitches can it sew in $\frac{1}{6}$ of a minute?
   (A) 2
   (B) 20
   (C) 30
   (D) 60
   (E) 720

GO ON TO THE NEXT PAGE.

5. 5,509 ÷ 7 =
   (A) 797
   (B) 787
   (C) 777
   (D) 767
   (E) 757

6. The average test score of three students was 78. If one of the boys did an extra credit assignment that brought his score up by 6 points and the other scores stayed the same, what will their average score be?
   (A) 28
   (B) 32
   (C) 80
   (D) 84
   (E) 96

7. Raul had 3 dozen roses in his shop. He sold $\frac{1}{9}$ of his 3 dozen roses each day for 4 days. How many roses does he have left?
   (A) 5
   (B) 8
   (C) 12
   (D) 16
   (E) 20

8. The radius of a circle is 9 inches. Using π = 3.14, what is the approximate circumference of the circle? *(Note: C = πd)*
   (A) 18 in
   (B) 28 in
   (C) 57 in
   (D) 64 in
   (E) 254 in

9. A cube has a side length of 3 feet. What is its volume?
   (A) 9 ft^3
   (B) 12 ft^3
   (C) 24 ft^3
   (D) 27 ft^3
   (E) 36 ft^3

10. Solve for z: −4z − (−17) = −3
    (A) −5
    (B) $-\frac{17}{7}$
    (C) $-\frac{7}{2}$
    (D) $\frac{17}{7}$
    (E) 5

GO ON TO THE NEXT PAGE.

11. The radius of the circle on the right is n. A second circle (not shown) has an area 2.25 times as great as that of the circle on the right. Find the radius of the second circle, in terms of n.
    (A) $\frac{n}{3}$
    (B) $\frac{n}{2}$
    (C) $\frac{2n}{3}$
    (D) $\frac{3n}{2}$
    (E) $2n$

12. If 450 students were surveyed, how many more students take the train to school than ride the bus?
    (A) 162
    (B) 90
    (C) 72
    (D) 18
    (E) 6

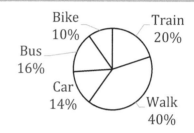

13. Which expression has a value that is 100 times the value of $50.8 \div 100$?
    (A) 508×0.1
    (B) 508×10
    (C) 50.8×0.1
    (D) 50.8×10
    (E) 5.08×100

14. A teacher purchases two boxes of cookies for a birthday party. Each box contains 34 cookies. If each student receives 3 cookies, and there are 2 cookies left over, how many students are in the class?
    (A) 10
    (B) 11
    (C) 22
    (D) 23
    (E) 32

15. $14 \times 16 \times 75 =$
    (A) 16,800
    (B) 8,400
    (C) 6,720
    (D) 4,200
    (E) 1,680

16. Which is the value of k in the equation $2^3 + k = 20$?
    (A) 8
    (B) 12
    (C) 14
    (D) 22
    (E) 28

GO ON TO THE NEXT PAGE.

17. If w more than 13 is 6 times a whole number, the value of w CANNOT be:
    (A) 5
    (B) 11
    (C) 13
    (D) 17
    (E) 23

18. The triangle shown is an isosceles triangle with AB = BC. What is the measure of angle BAC?
    (A) 156°
    (B) 132°
    (C) 78°
    (D) 48°
    (E) 24°

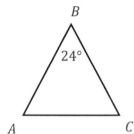

19. Which of the following is a prime number?
    (A) 67
    (B) 69
    (C) 72
    (D) 75
    (E) 77

20. Find the difference 6.38 – 0.0027.
    (A) 6.11
    (B) 6.353
    (C) 6.3773
    (D) 6.3793
    (E) 6.37973

21. Which fraction is equivalent to 190%?
    (A) $\frac{19}{100}$
    (B) $\frac{190}{100}$
    (C) $\frac{100}{190}$
    (D) $1\frac{9}{10}$
    (E) $1\frac{9}{100}$

22. About how many centimeters are in 89.74 inches?
    (Note: 1 inch = 2.54 centimeters)
    (A) 100 cm
    (B) 175 cm
    (C) 200 cm
    (D) 400 cm
    (E) 300 cm

GO ON TO THE NEXT PAGE.

23. Hunter is taking an exam and randomly guesses on the last 3 questions. If each question is multiple choice with 5 answer choices, what is the probability that he gets *all* 3 correct?
    (A) $\frac{4}{5}$
    (B) $\frac{3}{5}$
    (C) $\frac{1}{5}$
    (D) $\frac{1}{25}$
    (E) $\frac{1}{125}$

24. Meerko is a small business that was founded in 1995 and benefited from the growth of the Internet. If Meerko spends $5 million every year, during which periods did Meerko earn more than it spent?
    (A) 1997-2001 only
    (B) 1997-1999 and 2001-2004
    (C) 1998 and 2002-2004
    (D) 2001-2004 only
    (E) 1995-1997 and 1999-2001

Use the graph to the right to answer the following question.

25. What percent of Fort Riley residents are 21 – 40 or 81 – 100 years old?
    (A) 10
    (B) 20
    (C) 25
    (D) 30
    (E) 35

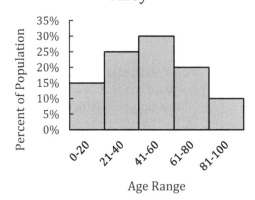

STOP
IF YOU FINISH BEFORE TIME IS UP,
CHECK YOUR WORK IN THIS SECTION ONLY.
YOU MAY NOT TURN TO ANY OTHER SECTION.

SECTION 2
40 Questions

Carefully read each passage and then answer the questions about it. For each question, select the choice that best answers the question based on the passage.

> Now was not the time to be distracted by the students sitting next to her. The teacher had just distributed the next set of test questions, and Willow picked up her copy eagerly. Her heart sank as she read it over. English history was not her best subject, and the facts she needed to know were precisely those which she could not remember. The
> 5 dread of failure was a fog that clouded her mind. She could only attempt about half of the questions, and even in these she could not recall specific dates. Suddenly, Patrick's advice flashed into her mind. "Write romantically, and make your paper sound poetic." It seemed like a long shot and she feared that it wouldn't be satisfactory, but in such a desperate situation, anything was worth trying. Willow possessed a certain flair for writing essays.
> 10 Learning to write had been her favorite lesson at Miss Hamilton's. She collected her thoughts, and did the very best she could. Choosing question four, "Tell about the life of Lady Jane Grey," she proceeded to treat the subject as creatively as possible. The pathetic tragedy of the young Jane Grey had always appealed to her imagination, and she could not have had a better topic to write about had she been able to choose from all the characters
> 15 in her history book.
> "Whom the gods love die young," she began, and paused. It was a good introduction, she decided.

1. The sentence "The dread of failure was a fog that clouded her mind" (lines 4-5) is an example of
   (A) cliché
   (B) idiom
   (C) simile
   (D) hyperbole
   (E) metaphor

2. According to the passage, Willow fears she will do poorly on the test because she
   (A) hates English history
   (B) failed to pay attention in Miss Hamilton's class
   (C) cannot help being distracted by other students
   (D) is unable to remember crucial information
   (E) must listen to Patrick's advice

3. Which of the following best describes Willow's reaction to Patrick's advice?
   (A) resentment
   (B) giddiness
   (C) animosity
   (D) skepticism
   (E) admiration

4. Over the course of the passage, Willow's feelings shift from
   (A) suspicion to hatred
   (B) excitement to bliss
   (C) despair to hopefulness
   (D) eagerness to disappointment
   (E) expectation to anger

5. Why does Willow choose to write about Lady Jane Grey?
   (A) Willow is similar to Lady Jane Grey.
   (B) Lady Jane Grey led a satisfactory life.
   (C) Lady Jane Grey was Patrick and Willow's favorite artist.
   (D) Willow can write most creatively about Lady Jane Grey.
   (E) Lady Jane Grey is the only historical figure Willow remembers.

GO ON TO THE NEXT PAGE.

In many ways, the Harlem Renaissance began as a dream looking for a creative outlet. African-Americans fleeing the racism of the South found a welcome haven in Harlem. In Harlem, skin color didn't prevent people from having a well-paying job or going out on the town. And in Harlem, people could write what they felt, what they
5 believed, and what they hoped for.
   During the 1920s and 30s, Harlem was the place where social forces, cultural pride and creative passion collided. That collision caused a truly remarkable artistic outpouring by African-American writers, artists and musicians.
   The writers of the Harlem Renaissance, whom we focus on today, celebrated their
10 culture in poetry and prose while capturing the stark realities of being black in America. In committing their words to paper, they shaped a rich literary history and stood witness to an important historical moment.
   Zora Neale Hurston captured the extremes of human emotion when she wrote, "I have been in sorrow's kitchen and licked out all the pots. Then I have stood on the peaky
15 mountain wrapped in rainbows, with a harp and a sword in my hands."
   Langston Hughes lamented the inequalities around him in "I, Too, Sing America"; and in "Harlem (2)," he asked, "What happens to a dream deferred?"
   The words of these writers opened us to the truth at a time when it most needed to be heard.

6. The author's purpose in writing this text is to
   (A) persuade the government to change laws
   (B) convince readers to avoid becoming artists
   (C) inform about accomplishments during a specific historical moment
   (D) prove that literature can cause change
   (E) detail the major events in the lives of famous African American artists

7. The word "stark" (line 10) most likely means
   (A) harsh
   (B) complex
   (C) untold
   (D) wonderful
   (E) rebellious

8. The phrase "I have been in sorrow's kitchen and licked out all the pots" (lines 13-14) suggests that the speaker
   (A) was ignorant and lacked manners
   (B) had a big appetite and loved food
   (C) worked in a kitchen despite not wanting to
   (D) disliked living in Harlem during the Harlem Renaissance
   (E) had experienced countless forms of negative treatment

9. According to the passage, why did Harlem become such a center of cultural expression?
   (A) It was where many artistic geniuses were born.
   (B) African American writers and artists faced racism in other parts of the country.
   (C) African American writers and artists enjoyed going out on the town there.
   (D) It was an inspirational place in the 1920s and 30s.
   (E) It was the place where Langston Hughes wrote "I, Too, Sing America"

10. The author includes information about Langston Hughes in order to
    (A) illustrate a point of view different from Zora Neale Hurston's
    (B) show how artists were able to capture and share feelings
    (C) explain why the Harlem Renaissance faded away
    (D) give information about why there was inequality during the 1920s
    (E) document his friendship with Zora Neale Hurston

> These wet rocks where the tide has been,
>   Barnacled white and weeded brown
> And slimed beneath to a beautiful green,
>   These wet rocks where the tide went down
> 5 Will show again when the tide is high
>   Faint and perilous, far from shore,
> No place to dream, but a place to die,—
>   The bottom of the sea once more.
> There was a child that wandered through
> 10   A giant's empty house all day,—
> House full of wonderful things and new,
>   But no fit place for a child to play.

11. The author's tone throughout the poem is best described as
    (A) comical
    (B) angry
    (C) cautious
    (D) tired
    (E) playful

12. The setting of the poem is most likely a
    (A) desert
    (B) ferry
    (C) field
    (D) beach
    (E) mountaintop

13. The author of the passage seems to be most interested in
    (A) sharing stories about giants
    (B) illustrating the appearance of the tides
    (C) describing the power and effects of the tides
    (D) recommending a safe place for children to play
    (E) discussing the lifecycle of different aquatic life

14. The author uses all of the following to describe the "wet rocks" (lines 1 and 4) EXCEPT
    (A) "barnacled white" (line 2)
    (B) "weeded brown" (line 2)
    (C) "beautiful green" (line 3)
    (D) "faint and perilous" (line 6)
    (E) "no fit place" (line 12)

15. The author would most likely agree that when the tide is high,
    (A) children should play on the rocks
    (B) the rocks become a place of dreams
    (C) the rocks become a beautiful green color
    (D) children can use the rocks to build a giant house
    (E) a fun place becomes dangerous instead

GO ON TO THE NEXT PAGE.

Today, some six billion people have access to a cell phone. For some, this is a cause for concern. Some people believe cell phones are distracting and addicting. Others say that cell phones cause cancer. People certainly spend more time on their cell phones than they should. But, use of these devices has not been shown to result in cancer.

5   Those who believe this rumor blame radiation. Overexposure to some types of radiation can lead to cancer. Medical x-rays, for instance, can be harmful in large doses. This is because these types of radiation are ionizing. This means they contain high-energy rays that can cause burns, sickness, and even cancer. Cell phones, on the other hand, emit non-ionizing rays. These rays are not powerful enough to be harmful to humans.

10 Microwave ovens and radios produce rays of similar strength. Yet, these devices are considered safe.

If cell phones were truly dangerous, there would be evidence. In fact, we find the opposite is true. Cell phone usage continues to rise but cancer rates do not. Cell phone usage increased dramatically between 1987 and 2005. During the same period, there was

15 no similar increase in brain cancer or cancer of the nervous system. Furthermore, government scientists have spent many hours testing cell phones. The conclusion is always the same: cell phone radiation is not dangerous.

16. The headline that best fits the article is
    (A) Avoid Medical X-Rays!
    (B) Scientists Discover New X-Rays
    (C) Cell Phones are Too Distracting
    (D) Number of Cell Phones Increasing
    (E) No Link Between Cell Phones & Cancer

17. Without changing the author's meaning, "dramatically" in line 14 could be replaced by
    (A) theatrically
    (B) minimally
    (C) substantially
    (D) pleasantly
    (E) loudly

18. The author's attitude towards the idea that cell phones causes cancer is one of
    (A) mild suspicion
    (B) lack of interest
    (C) complete acceptance
    (D) courteous disagreement
    (E) total understanding

19. The author implies which of the following about trends between 1987 and 2005 (lines 13-15)?
    (A) They are still being analyzed.
    (B) They disprove the author's opinion.
    (C) They contradict studies by the government.
    (D) They are evidence that cell phone usage is decreasing.
    (E) They support the claim that cell phones are harmless.

20. If radon gas is a source of ionizing radiation, then it must be
    (A) less dangerous than a radio
    (B) more dangerous than cell phones
    (C) less dangerous than a microwave oven
    (D) more dangerous than medical x-rays
    (E) very distracting and addicting to people

> "I once knew a little boy whose father and mother died when he was six years old. He was a slave and had no one to care for him. He slept on a dirt floor near a stove...In cold weather, he would crawl into a grain bag, headfirst, and leave his feet in the ashes to keep warm. Often, he would roast an ear of corn and eat it to satisfy his hunger. Many
> 5 times, he had to crawl under the barn or stable to secure some eggs, which he would roast in the fire and eat.
>
> This boy did not wear nice clothes as you do – only a linen shirt. Schools were unknown to him, and he learned to spell from an old Webster's spelling book. He learned to read and write from posters on cellars and barn doors. Eventually, he began to preach
> 10 and speak to others like him, and soon became well-known across the nation. He eventually held several high positions in society and accumulated a decent amount of wealth. He wore suits and no longer had to divide crumbs with the dogs under the table.
>
> That boy was me.
>
> What was possible for me is possible for you. Strive earnestly to add to your
> 15 knowledge. So long as you remain in ignorance, so long you will fail to command the respect of your fellow man."

21. What is the main idea of Frederick Douglass's speech?
    (A) Knowledge can lead to a better life.
    (B) Ignorance is bliss.
    (C) Merit is a myth created by fellow men.
    (D) Hard work is not necessary for success.
    (E) Education should be free for everyone.

22. The word "secure" (line 5) most likely means
    (A) protect
    (B) fortify
    (C) fasten or lock
    (D) seize
    (E) consume

23. According to the speech, Frederick Douglass's early education can be best described as
    (A) self-directed
    (B) private
    (C) tedious
    (D) expensive
    (E) disciplined

24. How does the phrase "no longer had to divide crumbs with the dogs under the table" (line 12) affect the mood of the story?
    (A) It shows a sense of community in the story, between Douglass and his pets.
    (B) It reinforces the hopefulness of the story by showing how even extreme poverty can be overcome.
    (C) It increases the despair of the story by evoking a sad moment in Frederick Douglass's childhood.
    (D) It emphasizes resentment in the story by discussing a moment of unfair deprivation in Frederick Douglass's childhood.
    (E) It shows the cheerfulness of the story by focusing on Douglass's success.

25. Which adjectives best describes the boy in the story?
    (A) wealthy and well-known
    (B) intelligent and resourceful
    (C) orphaned and depressed
    (D) earnest and ignorant
    (E) determined and honest

GO ON TO THE NEXT PAGE.

Hunting whales was big business during the nineteenth century. In 1846, the United States had 640 whaling ships, more than three times the rest of the world's ships combined. What made people hunt these giants of the sea to near extinction?

At that time, many parts of a whale were useful to people. Baleen is a stiff but
5   flexible material that some whales use to strain water out of their mouths when they feed. People used this material to make things like fishing rods and umbrellas. And ambergris, a curiously scented substance that one kind of whale, the sperm whale, produces in its stomach, was prized as an ingredient in perfumes. But most important of all was the whale's oil. It could be used to make candles, as fuel for lamps, or as a lubricant to keep
10   machine parts running smoothly. Sperm whales produce oil of very high quality in an organ in their heads. Scientists think this oil-filled organ helps sperm whales focus the sound waves that they use to find prey. Sperm oil was the most highly prized oil because it burned brightly and produced little smoke. Whale blubber, the layer of fat that all species of whales have to keep them warm, could also be used to make oil, but it was less
15   valuable than sperm oil because it could spoil in storage.

By the beginning of the twentieth century, however, whaling in the United States had gone the way of the whales. Fossil fuels like coal and oil, which formed from the buried remains of long dead plants and animals, had become easier to extract from the ground and use as a source of energy. Factories were being built across the country, and
20   fossil fuels like petroleum and natural gas could fuel America's booming industries more cheaply and effectively than whale oil could. These factories had also begun to draw workers away from jobs on whaling ships. Sending whale ships out onto the seas had gotten too expensive. Whaling just didn't pay off the way it once did.

26. As used in the sentence, the word "extract" (line 18) means
(A) find in.
(B) make from.
(C) create with.
(D) pull out.
(E) improve upon.

27. The best title for this passage is
(A) The Rise and Fall of US Whaling
(B) Why Do People Hunt Whales?
(C) The Whale Oil Rush
(D) The Uses of the Whale
(E) How the Whales Were Saved

28. According to the passage, the most important reason that whaling in the US declined was that
(A) whale blubber could spoil.
(B) whales were close to extinction.
(C) whaling had become too dangerous.
(D) whalers could not supply enough whale oil.
(E) fossil fuels had begun to replace whale oil.

29. The author uses the phrase "gone the way of the whales" (line 17) to suggest that
(A) whales were dying out.
(B) hunting whales was illegal.
(C) whale populations were growing.
(D) whale fossils were increasingly valuable.
(E) whale bones were a new source of materials.

30. How does the second paragraph support the passage's main idea?
(A) It tells how whaling started in the United States.
(B) It explains why hunting whales is wrong.
(C) It describes how whale oil was not as effective as fossil fuel.
(D) It describes the uses of whale parts to explain why the whaling industry had become so successful.
(E) It describes the uses of whale parts to help show how people lived in the nineteenth century.

GO ON TO THE NEXT PAGE.

> The work was difficult because the mines were hotter than the sun. It was not always unpleasant, however, and some of the miners would work in the mines throughout the night. But because sunlight never permeated the mines, it was difficult to tell if it was night or day down there.
>
> 5   Some miners who worked during the night would hear strange noises in the dark, even though they knew they were alone. They would declare the next morning that they heard a "tap-tapping" every time they stopped to take a breath. It sounded like the mountain was more full of miners at night than it ever was during the day. Many miners knew that the "tap-tapping" was the sound of goblins and refused to stay overnight. The
>
> 10 more courageous of the miners, however, like Peter Peterson and Curdie, stayed in the mine all night again and again.
>
>   Although they had encountered a few stray goblins several times, they never failed at driving them away with songs. It was well known that songs were the best defense against goblins. Goblins hated every kind of song, most likely because they did
>
> 15 not know how to sing any themselves. It is understandable that miners who could not make up songs nor remember old songs were truly afraid of the goblins. They were helpless! On the other hand, those who could make songs for themselves were never frightened. They knew that old songs were very effective at disturbing the goblins, and new songs even better at scaring them away.

31. In line 1, "hotter than the sun" is an example of
    (A) allusion
    (B) irony
    (C) simile
    (D) alliteration
    (E) hyperbole

32. As used in line 3, "permeated" most nearly means
    (A) struck
    (B) worked
    (C) disturbed
    (D) challenged
    (E) penetrated

33. It can be inferred from the passage that
    (A) miners do not usually work in the mines at night.
    (B) Peter Peterson and Curdie are the leaders of the miners.
    (C) the mine was rich in gold, making it worth it to work at night.
    (D) goblins were very dangerous, and often attacked the miners.
    (E) miners are very noisy, therefore disturbing and angering the goblins.

34. The passage suggests that goblins hate songs because they
    (A) interrupt their work.
    (B) cannot understand the songs.
    (C) have trouble hearing at night.
    (D) echo loudly throughout the mines
    (E) are unable to sing songs of their own.

35. According to the passage, goblins would most likely terrify
    (A) Curdie.
    (B) Peter Peterson.
    (C) a miner who has trouble memorizing songs.
    (D) a miner only familiar with older songs.
    (E) a creative miner who can come up with his own songs.

GO ON TO THE NEXT PAGE.

Abraham Lincoln's impressive speaking skills helped make him the President of the United States. More than 150 years after his assassination, people still honor his words and acknowledge his skill as an orator.

Just what made Lincoln such an effective speaker? How did a man from humble
5   beginnings go on to deliver the Gettysburg Address, a speech which the politician Robert G. Ingersoll believed would "live until languages are dead and lips are dust?"

The answer: deliberate simplicity. Lincoln once said, "Don't shoot too high. Aim low and the common people will understand you."

Lincoln's sentences were usually short. He would say, for example, "I dug a ditch,"
10   instead of, "I excavated a channel." Lincoln didn't just keep sentences short – he kept words short, as well. By some accounts, more than fifty percent of the words used in his speeches are just one syllable long. For example, 200 of the 271 words in the Gettysburg address are monosyllabic.

It's not that Lincoln couldn't use more complicated sentences and bigger words;
15   it's that he didn't want to. As Lincoln used to say, "The educated and refined people will understand you, anyway. If you aim too high your ideas will go over the heads of the masses and only hit those who need no hitting."

36. The tone of the passage is one of
    (A) joy
    (B) admiration
    (C) disapproval
    (D) frustration
    (E) precaution

37. As it is used in line 3, the word "orator" most likely refers to a(n)
    (A) honest person
    (B) powerful person
    (C) victim of a crime
    (D) public speaker
    (E) talented writer

38. The passage is mostly about
    (A) the Gettysburg Address
    (B) Abraham Lincoln's speaking skills
    (C) Abraham Lincoln's most famous sayings
    (D) the events of Abraham Lincoln's presidency
    (E) Robert G. Ingersoll's reflections on Abraham Lincoln

39. In line 13, "monosyllabic" most nearly means
    (A) boring
    (B) short
    (C) intentional
    (D) complicated
    (E) conventional

40. The author uses the phrase "'I dug a ditch,' instead of, 'I excavated a channel'" (lines 9 and 10) to highlight
    (A) the importance of the Gettysburg Address
    (B) Robert G. Ingersoll's view of Abraham Lincoln
    (C) Abraham Lincoln's political beliefs
    (D) Abraham Lincoln's simple way of speaking
    (E) Abraham Lincoln's last words before his assassination

STOP
IF YOU FINISH BEFORE TIME IS UP,
CHECK YOUR WORK IN THIS SECTION ONLY.
YOU MAY NOT TURN TO ANY OTHER SECTION.

SECTION 3
60 Questions

There are two different types of questions in this section: synonyms and analogies. Read the directions and sample question for each type.

Synonyms

Each of the questions that follow consist of one capitalized word. Each word is followed by five words or phrases. Select the one word or phrase whose meaning is closest to the word in capital letters.

Sample Question:

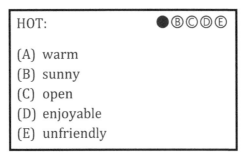

1. OMIT:
   (A) leave out
   (B) make sure
   (C) perform well
   (D) send around
   (E) pull through

2. PRIME:
   (A) fattest
   (B) automobile
   (C) number
   (D) best
   (E) indivisible

3. DESTRUCTIVE:
   (A) interpret
   (B) demolish
   (C) harmful
   (D) civil
   (E) monstrous

4. ENCLOSURE:
   (A) zone
   (B) assault
   (C) stop
   (D) surround
   (E) fencing

5. CREATE:
   (A) copy
   (B) inventor
   (C) edit
   (D) product
   (E) make

6. TARGET:
   (A) aim
   (B) battle
   (C) acquire
   (D) arrow
   (E) store

7. FURTHERMORE:
   (A) distant
   (B) above
   (C) many
   (D) tally
   (E) additionally

8. REVIVE:
   (A) resurrect
   (B) append
   (C) alive
   (D) canine
   (E) immortal

GO ON TO THE NEXT PAGE.

9. DISPEL:
   (A) step into
   (B) drive away
   (C) send up
   (D) move ahead
   (E) count down

10. EXTENSIVE:
    (A) depth
    (B) thorough
    (C) shyly
    (D) elegant
    (E) categorize

11. POSITIVE:
    (A) obtain
    (B) beneficial
    (C) personality
    (D) outcome
    (E) quality

12. UNNECESSARY:
    (A) basic
    (B) crucial
    (C) key
    (D) important
    (E) superfluous

13. INJURE:
    (A) painful
    (B) bandage
    (C) hurt
    (D) unjust
    (E) rest

14. ENERGETIC:
    (A) extra
    (B) amazing
    (C) lively
    (D) coffee
    (E) adorable

15. SIGNIFICANT:
    (A) amplify
    (B) transmit
    (C) represent
    (D) similar
    (E) meaningful

16. MECHANISM:
    (A) reliable
    (B) factory
    (C) mechanic
    (D) metallic
    (E) process

17. PROCLAIM:
    (A) borrow
    (B) say
    (C) take
    (D) own
    (E) disrupt

18. ENABLE:
    (A) access
    (B) force
    (C) wander
    (D) claim
    (E) permit

19. MATURE:
    (A) qualified
    (B) allege
    (C) attractive
    (D) sophisticated
    (E) age

20. ENORMOUS:
    (A) elephant
    (B) vastly
    (C) giant
    (D) weird
    (E) normal

21. IMAGE:
    (A) mind
    (B) portrait
    (C) concrete
    (D) imagination
    (E) picture

22. DISTORT:
    (A) criminal
    (B) truth
    (C) transformer
    (D) warp
    (E) straighten

GO ON TO THE NEXT PAGE.

23. IMPATIENCE:
    (A) nervous
    (B) resource
    (C) restlessness
    (D) excitement
    (E) busily

24. IMPROBABLE:
    (A) unlikely
    (B) fraction
    (C) beneficiary
    (D) certain
    (E) problematic

25. FILE:
    (A) version
    (B) computer
    (C) envelope
    (D) email
    (E) categorize

26. AGONY:
    (A) disappearance
    (B) duration
    (C) experience
    (D) suffering
    (E) remedy

27. SUPPLEMENT:
    (A) reduce
    (B) add
    (C) soften
    (D) potential
    (E) repair

28. EVENTUAL:
    (A) ended
    (B) ultimate
    (C) lasting
    (D) eventful
    (E) perpetual

29. DEFINITE:
    (A) purpose
    (B) select
    (C) exact
    (D) definition
    (E) minimal

30. EXCLUSIVE:
    (A) moderate
    (B) divide
    (C) concede
    (D) challenging
    (E) restricted

GO ON TO THE NEXT PAGE.

Analogies

The questions that follow ask you to find relationships between words. For each question, select the answer choice that best completes the meaning of the sentence.

Sample Question:

> Dance is to dancer as: ●ⒷⒸⒹⒺ
>
> (A) lesson is to teacher
> (B) cat is to yarn
> (C) fish is to water
> (D) umbrella is to rain
> (E) shovel is to snow

Choice (A) is the best answer because a dancer dances a dance, just as a teacher teaches a lesson. This choice states a relationship that is most like the relationship between dance and dancer.

---

31. Blizzard is to flurry as
    (A) bridge is to river
    (B) hurricane is to tornado
    (C) savior is to saint
    (D) skeletal is to thin
    (E) intensify is to weaken

32. Narrate is to tale as
    (A) emotion is to feeling
    (B) borrow is to lend
    (C) defraud is to steal
    (D) traffic is to direct
    (E) compete is to race

33. Stick is to puck as
    (A) job is to business
    (B) fiber is to wool
    (C) net is to basket
    (D) referee is to umpire
    (E) glove is to ball

34. Cheapest is to cheap as
    (A) quiet is to quietest
    (B) quicker is to quickest
    (C) nearest is to near
    (D) type is to typed
    (E) extreme is to extremely

35. Deliberately is to unintentionally as
    (A) ridiculously is to solitarily
    (B) eventful is to fun
    (C) confide is to warmly
    (D) sneak is to cheat
    (E) gracefully is to clumsily

36. Handsaw is to cut as
    (A) heel is to calf
    (B) belief is to ignore
    (C) prejudice is to segregation
    (D) gush is to flow
    (E) prank is to dupe

37. Person is to audience as
    (A) act is to play
    (B) stage is to recite
    (C) applause is to clap
    (D) auditorium is to director
    (E) celebration is to encore

38. Prominent is to notable as
    (A) neural is to anxious
    (B) safe is to insecure
    (C) impede is to help
    (D) fervid is to illness
    (E) disclose is to reveal

GO ON TO THE NEXT PAGE.

39. Miniscule is to gigantic as
    (A) joke is to mock
    (B) fertile is to barren
    (C) endure is to pressure
    (D) specify is to particular
    (E) succeed is to fumble

40. Detective is to culprit as
    (A) amusement is to park
    (B) train is to track
    (C) rooster is to hen
    (D) dog is to ball
    (E) basket is to handle

41. Circus is to tent as painting is to
    (A) frame
    (B) still
    (C) bronze
    (D) sculpt
    (E) liberty

42. Book is to dictionary as
    (A) kin is to mother
    (B) library is to edition
    (C) daughter is to son
    (D) carrot is to vegetable
    (E) fact is to fiction

43. Potter is to clay as
    (A) librarian is to referenced
    (B) janitor is to mop
    (C) construct is to plank
    (D) shoemaker is to cobbler
    (E) publicist is to famous

44. Learning is to wisdom as
    (A) heat is to burn
    (B) crease is to wrap
    (C) famine is to hungry
    (D) dedication is to memory
    (E) region is to location

45. Harvest is to crop as
    (A) caress is to smoothly
    (B) address is to area
    (C) curious is to questionable
    (D) resolve is to conflict
    (E) religion is to preach

46. Droplet is to ocean as rock is to
    (A) sediment
    (B) mountain
    (C) pebble
    (D) erode
    (E) jagged

47. Jester is to amuse as
    (A) eavesdrop is to listen
    (B) pyramid is to pointy
    (C) queen is to reign
    (D) rent is to mortgage
    (E) collect is to debt

48. Clothing is to sweater as cutlery is to
    (A) utensil
    (B) edible
    (C) metallic
    (D) spoon
    (E) sharp

49. Waiter is to tray as
    (A) owner is to supervise
    (B) clandestine is to secretive
    (C) video is to film
    (D) timepiece is to hour
    (E) writer is to notebook

50. Often is to frequently as
    (A) ocular is to sight
    (B) similar is to different
    (C) obtuse is to angle
    (D) torpid is to energetic
    (E) horde is to mob

GO ON TO THE NEXT PAGE.

51. Infinite is to limited as reluctant is to
    (A) partner
    (B) hesitant
    (C) cautious
    (D) eager
    (E) regretful

52. Bad is to worse as
    (A) fix is to fixed
    (B) fast is to faster
    (C) greedily is to greedier
    (D) easier is to easy
    (E) argue is to arguably

53. Congratulate is to winner as consider is to
    (A) reasonable
    (B) considerate
    (C) excellent
    (D) optimal
    (E) choice

54. Philosopher is to theory as
    (A) bomb is to damage
    (B) echo is to ear
    (C) fear is to afraid
    (D) swiftly is to barely
    (E) decorate is to beauty

55. Fish is to net as
    (A) dollar is to currency
    (B) depress is to deflate
    (C) modifier is to change
    (D) picture is to paintbrush
    (E) pride is to honor

56. Gap is to abyss as
    (A) route is to continuous
    (B) slick is to slippery
    (C) cavern is to explorer
    (D) stumble is to foot
    (E) cool is to freeze

57. House is to bedroom as
    (A) estate is to buyer
    (B) schedule is to appointment
    (C) chart is to office
    (D) week is to day
    (E) wallpaper is to ceiling

58. Shape is to octagon as
    (A) rot is to decay
    (B) stop is to sign
    (C) wear is to adorn
    (D) chocolate is to chip
    (E) jewelry is to bracelet

59. Journalist is to news as
    (A) canvas is to surveillance
    (B) newspaper is to magazine
    (C) meteorologist is to weather
    (D) station is to channel
    (E) claim is to evidence

60. Tooth is to chew as theater is to
    (A) actress
    (B) ticket
    (C) craft
    (D) entertain
    (E) line

STOP
IF YOU FINISH BEFORE TIME IS UP,
CHECK YOUR WORK IN THIS SECTION ONLY.
YOU MAY NOT TURN TO ANY OTHER SECTION.

SECTION 4
25 Questions

There are five suggested answers after each problem in this section. Solve each problem in your head or in the space provided to the right of the problem. Then look at the suggested answers and pick the best one.

Note: Any figures or shapes that accompany problems in Section 1 are drawn as accurately as possible EXCEPT when it is stated that the figure is NOT drawn to scale.

Sample Question:

11 + 13 =   ●ⒷⒸⒹⒺ
    (A) 24
    (B) 113
    (C) 131
    (D) 241
    (E) 311

DO WORK IN THIS SPACE

1. Ramona is making juice from a powder. The recipe calls for $\frac{1}{4}$ cup of sugar, but Ramona wants to use $\frac{1}{3}$ of that amount. How many cups should she use?
   (A) $\frac{1}{12}$
   (B) $\frac{1}{7}$
   (C) $\frac{2}{7}$
   (D) $\frac{1}{3}$
   (E) $\frac{3}{4}$

2. Which expression represents 11 more than three-fourths of x?
   (A) $11 - \frac{3}{4}x$
   (B) $11(\frac{3}{4}) + x$
   (C) $\frac{3}{4}x + 11$
   (D) $\frac{3}{x} + 11$
   (E) $\frac{3}{4}x - 11$

3. The ratio of grapefruits to lemons to pomegranates in a fruit basket is 3:6:2. The total number of grapefruits and pomegranates combined is 25. How many lemons are there?
   (A) 10
   (B) 15
   (C) 20
   (D) 30
   (E) 55

GO ON TO THE NEXT PAGE.

4. Malcolm spent $1\frac{1}{2}$ hours on homework on Monday. He spent 75 minutes on homework on Tuesday. How much longer did Malcolm work on homework on Monday than Tuesday?
   (A) 15 minutes
   (B) 30 minutes
   (C) 45 minutes
   (D) 90 minutes
   (E) 105 minutes

5. A point at (–3, –4) is reflected across the *x*-axis. What is its new location?
   (A) (–3, 4)
   (B) (3, 4)
   (C) (3, –4)
   (D) (–4, 3)
   (E) (4, –3)

6. Christy has $25 dollars. If popsicles cost $1.20 each, what is the greatest number of popsicles she can buy with her money?
   (A) 18
   (B) 19
   (C) 20
   (D) 21
   (E) 22

7. 37 × 2,997 =
   (A) 119,889
   (B) 111,789
   (C) 110,899
   (D) 110,889
   (E) 100,089

8. A triangle has an interior angle that measures greater than 90°. Which of the following terms could describe the triangle?
   (A) equilateral
   (B) right
   (C) right-isosceles
   (D) acute
   (E) isosceles

GO ON TO THE NEXT PAGE.

9. Priya works to earn money for a new phone. The number of hours she worked each week for the first six weeks is displayed on the right.

   Between which two weeks did the number of hours Priya worked increase 50%?
   (A) weeks 1 and 2
   (B) weeks 2 and 3
   (C) weeks 3 and 4
   (D) weeks 4 and 5
   (E) weeks 5 and 6

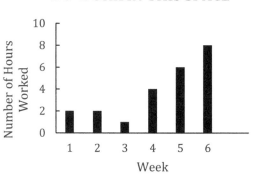

10. Polygons *LMNP* and *QRST* are both squares. What is the area of the shaded region?
    (A) 14 m²
    (B) 28 m²
    (C) 32 m²
    (D) 38 m²
    (E) 48 m²

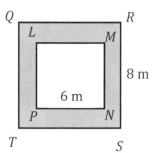

11. In the series below, $\frac{2}{3}$ is the first term. For each term after the first, the numerator is one more than the numerator in the preceding term and the denominator is two more than the denominator in the preceding term. What is the value of *x*?

    $$\frac{2}{3}, \frac{3}{5}, \frac{4}{7}, \frac{5}{9}, \frac{6}{11}, \ldots, \frac{x}{19}$$

    (A) 8
    (B) 9
    (C) 10
    (D) 11
    (E) 12

12. A six sided number cube has sides labeled 1-6. What is the probability of rolling a number higher than 4?
    (A) $\frac{1}{6}$
    (B) $\frac{1}{5}$
    (C) $\frac{1}{4}$
    (D) $\frac{1}{3}$
    (E) $\frac{1}{2}$

GO ON TO THE NEXT PAGE.

13. How many positive integer factors does 75 have?
    (A) 3
    (B) 4
    (C) 6
    (D) 7
    (E) 10

14. Chris is stringing beads together in the pattern one blue, one yellow, one orange, one red, one violet. What color will the 54th term be?
    (A) Blue
    (B) Yellow
    (C) Orange
    (D) Red
    (E) Violet

15. What is the radius of a circle whose circumference is 64π units?
    (Note: C = πd)
    (A) 4
    (B) 8
    (C) 32
    (D) 64
    (E) 201

16. Tyler completes 8 math problems in 5 minutes. At that rate, how many math problems will he complete in one hour?
    (A) 37
    (B) 40
    (C) 48
    (D) 80
    (E) 96

17. To keep the 7 camels in a zoo healthy, 25 liters of water are needed per week. How many camels could survive on 100 liters of water for a week?
    (A) 26
    (B) 27
    (C) 28
    (D) 29
    (E) 30

18. Kori has a box of 30 new pencils. She sharpens $\frac{2}{5}$ of the pencils. How many did she leave unsharpened?
    (A) 8
    (B) 10
    (C) 12
    (D) 15
    (E) 18

19. Principal Lopez will order textbooks that are shipped in boxes for 7 classes, each with 25 students. Each box contains up to 15 textbooks. If each student receives one textbook, how many boxes should Principal Lopez order?
    (A) 5
    (B) 11
    (C) 12
    (D) 75
    (E) 175

20. An artist runs a table at a craft show one weekend. At the end of the weekend, she calculates she made a total of $123.46 in sales. She spent $24.78 on beads, $15.11 on string, and $34.89 on paint. Which of the following is the best estimate of the artist's profit?
    (A) $25
    (B) $50
    (C) $75
    (D) $100
    (E) $125

21. 149 × 604 =
    (A) 90,016
    (B) 90,004
    (C) 89,996
    (D) 89,884
    (E) 49,002

22. Which of the following is greater than $\frac{1}{4}$?
    (A) $\frac{3}{20}$
    (B) $\frac{1}{5}$
    (C) $\frac{2}{10}$
    (D) $\frac{2}{9}$
    (E) $\frac{1}{3}$

23. Which expression is equivalent to $12x + 6y$?
    (A) $6(2x + y)$
    (B) $6(2x + 6y)$
    (C) $12(x + y)$
    (D) $\frac{1}{2}(6x + 3y)$
    (E) $2(6x + 6y)$

24. Camila ordered pizzas for her party; there are 64 slices in total. Of these, 16 are veggie slices, and 32 are pepperoni. Which fraction of the pizzas is neither veggie nor pepperoni?
    (A) $\frac{1}{5}$
    (B) $\frac{1}{4}$
    (C) $\frac{1}{2}$
    (D) $\frac{3}{4}$
    (E) $\frac{7}{8}$

25. An integer is represented by the value K. What is the largest value of K when $K - 10 < -7?$
    (A) −17
    (B) −3
    (C) 2
    (D) 3
    (E) 4

STOP
IF YOU FINISH BEFORE TIME IS UP,
CHECK YOUR WORK IN THIS SECTION ONLY.
YOU MAY NOT TURN TO ANY OTHER SECTION.

SECTION 5
16 Questions

> Turmeric is processed from the bulbous root of the *Curcuma domestica* plant. It can be boiled, dried, and ground into a yellow-orange powder. It has a strong flavor and has been used in cooking for thousands of years.
>
> Today, many people also use turmeric for its purported health and beauty
> 5 benefits. Some Americans use the bitter powder as a replacement for costly pain pills. They believe it can help reduce swelling in the body. Others use turmeric to combat Alzheimer's disease. Still others believe that turmeric can help those with depression. Some even believe that turmeric improves the skin's appearance, and have added the spice into beauty creams.
> 10 However, there is little scientific evidence that supports any of these claims. A professional study conducted by turmeric enthusiasts at Swinburne University in Australia showed little scientific proof of the spice's curative effects. Early studies show turmeric *might* help with inflammation, but these studies were not carried out on humans. Until more research can be done, people would do well to use turmeric as our
> 15 ancestors have: as a flavorful cooking spice.

1. This passage is best described as
   (A) an encyclopedia entry
   (B) an argumentative essay
   (C) a fictional short story
   (D) a dramatic composition
   (E) a political speech

2. The final sentence (lines 14-15) is mainly used to
   (A) summarize how turmeric can help improve health
   (B) recap the scientific evidence behind turmeric's curative effects
   (C) discredit turmeric's medical properties
   (D) argue that people should use turmeric in their cooking
   (E) highlight the flavorful properties of turmeric

3. What is a good headline for this passage?
   (A) "Turmeric: A Medical Wonder"
   (B) "Turmeric Can Combat Depression and Alzheimer's"
   (C) "Turmeric: A Flavorful Spice"
   (D) "Turmeric: Food, Not Medicine"
   (E) "A History of Turmeric"

4. All of the following are listed as ways that people use turmeric EXCEPT as
   (A) a replacement for pain pills
   (B) a cure for Alzheimer's
   (C) a beauty product and skin cream
   (D) a digestive aid
   (E) an anti-depressant

5. The word "purported" (line 4) most likely means
   (A) supposed
   (B) definite
   (C) evidenced
   (D) discovered
   (E) fantastic

6. DISSOLVE:
   (A) disintegrate
   (B) faded
   (C) averse
   (D) solution
   (E) sugar

7. EXCLAMATION:
   (A) secure
   (B) mark
   (C) declare
   (D) shout
   (E) proportion

GO ON TO THE NEXT PAGE.

8. FEATURE:
   (A) personality
   (B) display
   (C) eyes
   (D) news
   (E) praise

9. Interested is to enthusiastic
   (A) furious is to angry
   (B) calm is to peaceful
   (C) incredulous is to disbelieving
   (D) hungry is to ravenous
   (E) suspicious is to untrustworthy

10. Freezing is to sneeze as hurricane is to
    (A) warmth
    (B) flooding
    (C) climate change
    (D) swimming
    (E) superstorm

11. Goose is to flock as hour is to
    (A) late
    (B) early
    (C) day
    (D) midnight
    (E) minute

12. Which fraction has a value that is nearest to 0.87?
    (A) $\frac{1}{87}$
    (B) $\frac{9}{10}$
    (C) $\frac{2}{3}$
    (D) $\frac{4}{7}$
    (E) $\frac{3}{4}$

13. Which inequality is equivalent to $\frac{3}{8} - v > 1$?
    (A) $v > -\frac{11}{8}$
    (B) $v < -\frac{5}{8}$
    (C) $v > -\frac{5}{8}$
    (D) $v < \frac{5}{8}$
    (E) $v > \frac{11}{8}$

14. If the diameter of a circle is 11 centimeters, what is the approximate area of the circle?
    (Note: $\pi = 3.14$, $A = \pi r^2$)
    (A) 17.27 cm²
    (B) 30.25 cm²
    (C) 34.54 cm²
    (D) 94.988 cm²
    (E) 379.94 cm²

15. Find the value of $0.2 - 0.3 \times 0.6$.
    (A) −0.06
    (B) 0.02
    (C) 0.06
    (D) 0.182
    (E) 0.3

16. Each of the numbers in the list below shows the number of hours Tom read over the course of 8 weeks. What is the mean number of hours Tom spent reading per week?
    8, 4, 6, 6, 2, 3, 7, 4
    (A) 4
    (B) 5
    (C) 6
    (D) 7
    (E) 8

STOP
IF YOU FINISH BEFORE TIME IS UP,
CHECK YOUR WORK IN THIS SECTION ONLY.
YOU MAY NOT TURN TO ANY OTHER SECTION.

Scoring the Final Practice Test (Form B)

Writing Sample – Unscored
Have a parent or trusted educator review the essay or story written for the writing sample. Important areas to focus on include organization, clarity of ideas, originality, and technical precision (spelling, grammar, etc.).

Sections 1-4 – Scored
Score the test using the answer sheet and *referring to the answer key in the back of the book (see table of contents)*.

Step 1: For each section, record the number of questions answered correctly.

Step 2: For each section, record the number of questions answered incorrectly. Then, multiply that number by ¼ to calculate the penalty.

Section	Questions Correct
Quantitative *Section 1 + Section 4*	————
Reading *Section 2*	————
Verbal *Section 3*	————

Section	Questions Incorrect	Penalty	
Quantitative *Section 1 + Section 4*	————	x $\frac{1}{4}$ =	————
Reading *Section 2*	————	x $\frac{1}{4}$ =	————
Verbal *Section 3*	————	x $\frac{1}{4}$ =	————

Step 3: For each section, subtract the Penalty in Step 2 from the Questions Correct in Step 1. This is the raw score. Note that the actual test will convert the raw score to a scaled score by comparing the student's performance with all other students in the same grade who took the test.

Section	Raw Score
Quantitative *Section 1 + Section 4*	————
Reading *Section 2*	————
Verbal *Section 3*	————

> **Consider**: How certain were you on the questions you guessed on? Should you have left those questions blank, instead? How should you change the way you guess and leave questions blank?

Carefully consider the results from the practice test when revising a study plan. Remember, the Middle Level SSAT is given to students in grades 5-7. Unless the student has finished 7th grade, chances are that there is material on this test that he or she has not yet been taught. If this is the case, and the student would like to improve beyond what is expected of his or her grade, consider working with a tutor or teacher, who can help learn more about new topics.

Section 5 – Unscored
On the real test, the Experimental section will NOT be scored. Consider the student's performance on this section for practice purposes only. Did he or she do better on one section than other? Use this information along with the information from Sections 1-4 to reevaluate the study plan.

Answer Keys

This section provides the answer solutions to the practice questions in each section of the workbook except for the practice tests and writing sample sections. The answers to the practice tests immediately follow their respective tests. There are no answers provided to the writing sample section. Instead, consider having a tutor, teacher, or other educator review your writing and give you constructive feedback.

Remember: detailed answer explanations are available online at **www.thetutorverse.com**. Students should ask a parent or guardian's permission before going online.

Diagnostic Practice Test (Form A) Answer Key

Section 1: Quantitative

1. E	5. B	9. C	13. D	17. B	21. C	25. D	
2. D	6. D	10. D	14. C	18. C	22. D		
3. C	7. C	11. A	15. E	19. D	23. C		
4. C	8. D	12. E	16. B	20. C	24. E		

Section 2 – Reading

1. A	6. C	11. B	16. B	21. B	26. B	31. C	36. D
2. D	7. D	12. B	17. D	22. C	27. C	32. E	37. D
3. B	8. A	13. E	18. A	23. B	28. B	33. B	38. C
4. E	9. C	14. D	19. E	24. C	29. D	34. A	39. B
5. E	10. A	15. B	20. D	25. D	30. D	35. B	40. D

Section 3 – Verbal

1. E	9. B	17. A	25. A	33. C	41. E	49. A	57. D
2. B	10. D	18. C	26. B	34. A	42. D	50. A	58. E
3. E	11. C	19. A	27. B	35. D	43. D	51. B	59. A
4. B	12. A	20. A	28. E	36. E	44. A	52. D	60. E
5. C	13. E	21. D	29. C	37. D	45. B	53. B	
6. E	14. A	22. C	30. A	38. C	46. E	54. D	
7. D	15. A	23. C	31. C	39. B	47. D	55. E	
8. E	16. E	24. C	32. C	40. C	48. E	56. E	

Section 4 – Quantitative

1. E	5. D	9. A	13. C	17. C	21. D	25. D	
2. B	6. E	10. B	14. C	18. D	22. C		
3. A	7. D	11. D	15. C	19. D	23. A		
4. E	8. A	12. D	16. D	20. C	24. C		

Section 5 – "Experimental"

1. E	3. E	5. A	7. A	9. A	11. B	13. E	15. C
2. D	4. C	6. E	8. C	10. E	12. D	14. E	16. D

Quantitative

Number Concepts & Operations

Place Value
1. E
2. B
3. D
4. C
5. D
6. E
7. A
8. C
9. B
10. E
11. D

Decimals
1. D
2. D
3. C
4. B
5. E
6. E
7. D
8. B
9. D
10. D
11. C
12. A

Fractions
1. E
2. B
3. E
4. D
5. C
6. A
7. A
8. C
9. B
10. E
11. C
12. D
13. C
14. D
15. C
16. B
17. D

Percents
1. B
2. C
3. D
4. B
5. E
6. B
7. D
8. B
9. C
10. D
11. D
12. C
13. E
14. E
15. D
16. B
17. B

Decimals/Fractions/Percents
1. C
2. B
3. C
4. C
5. D
6. E
7. A
8. A
9. E
10. D
11. D
12. B
13. C
14. C
15. D
16. D

Whole Numbers
1. B
2. D
3. D
4. B
5. C
6. C
7. C
8. B
9. B
10. D
11. B
12. B

Order of Operations
1. B
2. C
3. E
4. B
5. B
6. B
7. B
8. E
9. D
10. E
11. E
12. D

Pre-Algebra

Ratio and Proportions
1. B
2. D
3. D
4. C
5. B
6. D
7. D
8. E
9. C
10. B
11. C
12. C
13. E
14. E
15. D
16. B
17. C
18. B
19. C
20. B
21. C
22. B
23. B
24. E

Sequences, Patterns, Logic
1. A
2. E
3. C
4. D
5. B
6. B
7. E
8. E
9. E
10. D
11. E
12. E
13. D

Estimation
1. C
2. B
3. D
4. C
5. D
6. B
7. C
8. B
9. D
10. E
11. B

Algebra

Interpreting Variables
1. D
2. A
3. E
4. C
5. D
6. E
7. B
8. A
9. B
10. A
11. E
12. B
13. C
14. E
15. D
16. B
17. B
18. D
19. C
20. B
21. C
22. C

Solving Equations and Inequalities
1. C
2. B
3. D
4. B
5. E
6. D
7. B
8. E
9. A
10. E
11. B
12. C
13. E
14. E
15. B
16. C
17. A
18. A
19. B
20. D
21. D
22. E
23. C

Multi-Step Word Problems
1. D
2. C
3. B
4. B
5. B
6. A
7. A
8. E
9. A
10. A
11. C
12. B
13. B
14. E
15. C
16. C
17. C
18. C
19. D
20. E
21. A
22. D
23. C

Geometry

Pythagorean Theorem
1. B
2. C
3. C
4. A
5. B
6. C
7. B
8. D
9. D

Coordinate Plane
1. B
2. C
3. E
4. B
5. C
6. A
7. E
8. C
9. D
10. A
11. C

Transformations
1. A
2. C
3. A
4. C
5. A
6. E
7. E
8. B
9. B
10. D
11. B

Circles
1. C
2. B
3. C
4. E
5. D
6. D
7. E
8. E
9. B
10. E
11. A

Two and Three Dimensional Shapes
1. E
2. D
3. E
4. E
5. A
6. D
7. D
8. B
9. E
10. C

Spatial Reasoning
1. C
2. D
3. E
4. B
5. D
6. E

Measurement

Time and Money
1. C
2. D
3. C
4. C
5. D
6. D
7. D
8. C
9. C
10. C

Area, Perimeter and Volume
1. C
2. B
3. A
4. D
5. C
6. A
7. B
8. C
9. D
10. C
11. D
12. E
13. D
14. B
15. D
16. B
17. B

Angles
1. C
2. D
3. C
4. A
5. D
6. E
7. B
8. D
9. A
10. D
11. C
12. E
13. E

Unit Analysis
1. C
2. C
3. C
4. C
5. A
6. B
7. E
8. B
9. B
10. A

Data Analysis

Interpreting Bar Graphs
1. D
2. B
3. B
4. C
5. A
6. D
7. A
8. C
9. E
10. C
11. D

Interpreting Histograms
1. B
2. C
3. A
4. E
5. E
6. B
7. C
8. B
9. C
10. D

Interpreting Line Graphs
1. C
2. D
3. A
4. D
5. B
6. C
7. D
8. E
9. C
10. D

Interpreting Circle Graphs
1. C
2. E
3. B
4. C
5. E
6. D
7. C
8. D
9. B
10. D

Statistics & Probability

Basic Probability
1. A
2. D
3. C
4. A
5. C
6. B
7. C
8. D
9. A
10. D
11. B
12. C
13. E

Compound Events
1. C
2. A
3. E
4. A
5. A
6. C
7. B
8. E
9. D
10. B
11. D

Answer Keys

Mean, Median, Mode
1. E   3. B   5. C   7. A   9. B   11. B   13. B
2. C   4. D   6. E   8. A   10. D   12. D

Verbal – Synonyms

Introductory
1. D   6. B   11. B   16. C   21. D   26. E   31. C
2. E   7. D   12. E   17. B   22. C   27. D   32. E
3. E   8. D   13. A   18. C   23. A   28. D   33. B
4. A   9. E   14. E   19. D   24. E   29. D
5. D   10. B   15. A   20. E   25. A   30. A

Intermediate
1. D   6. B   11. D   16. E   21. B   26. D   31. A
2. A   7. C   12. B   17. D   22. B   27. A   32. B
3. C   8. D   13. D   18. B   23. B   28. A   33. A
4. E   9. C   14. D   19. E   24. D   29. A
5. C   10. C   15. E   20. A   25. D   30. C

Advanced
1. B   6. B   11. B   16. B   21. E   26. A   31. A
2. B   7. B   12. E   17. A   22. D   27. E   32. D
3. D   8. C   13. B   18. B   23. E   28. D   33. E
4. A   9. C   14. B   19. D   24. A   29. B
5. A   10. E   15. C   20. D   25. A   30. E

Verbal – Analogies

Guided Practice – Antonyms
1. B   2. D   3. E   4. D   5. A   6. D   7. C   8. A

Guided Practice – Cause-and-Effect
1. D   2. A   3. E   4. C   5. D   6. E   7. E   8. A

Guided Practice – Defining
1. C   2. A   3. E   4. D   5. A   6. B   7. C   8. E

Guided Practice – Degree/Intensity
1. E   2. C   3. D   4. A   5. B   6. D   7. B   8. C

Guided Practice – Function/Object
1. A   2. C   3. B   4. D   5. A   6. D   7. B   8. A

Guided Practice – Grammar
1. C   2. D   3. E   4. D   5. A   6. C   7. B   8. C

Guided Practice – Individual/Object
1. A   2. A   3. C   4. E   5. B   6. C   7. C   8. E

Guided Practice – Noun/Verb
1. C   2. E   3. D   4. E   5. E   6. E   7. A   8. A

Guided Practice – Part/Whole
1. C   2. C   3. B   4. E   5. D   6. C   7. B   8. A

Guided Practice – Purpose/Object
1. E   2. C   3. D   4. B   5. C   6. D   7. D   8. B

Guided Practice – Type/Kind
1. E   2. A   3. A   4. D   5. B   6. E   7. C   8. B

Guided Practice – Synonym
1. E   2. E   3. D   4. A   5. A   6. A   7. C   8. D

Mixed Practice
1. C   3. C   5. B   7. A   9. D   11. E
2. E   4. C   6. E   8. E   10. D   12. B

Reading Comprehension

Fiction

Passage #1
1. C   2. D   3. C   4. D   5. E

Passage #2
1. E   2. B   3. D   4. E   5. D

Passage #3
1. D   2. E   3. A   4. A   5. B

Passage #4
1. E   2. D   3. B   4. C   5. E

Passage #5
1. A   2. D   3. B   4. D   5. D

Passage #6
1. C   2. A   3. C   4. D   5. B

Passage #7
1. A   2. B   3. A   4. A   5. C

Passage #8
1. C   2. E   3. C   4. A   5. B

Passage #9
1. A   2. C   3. A   4. E   5. C

Passage #10
1. E   2. B   3. C   4. C   5. E

Nonfiction

Passage #1
1. B   2. D   3. A   4. D   5. B

Passage #2
1. D   2. B   3. B   4. D   5. C

Passage #3
1. A   2. B   3. A   4. B   5. B

Passage #4
1. B   2. A   3. E   4. B   5. E

Passage #5
1. A   2. B   3. C   4. A   5. C

Passage #6
1. E   2. E   3. B   4. C   5. D

Passage #7
1. B   2. E   3. D   4. E   5. E

Passage #8
1. E   2. E   3. C   4. B   5. C

Passage #9
1. D   2. E   3. C   4. E   5. E

Passage #10
1. B   2. E   3. D   4. D   5. D

Final Practice Test (Form B) Answer Key

Section 1: Quantitative

1. A	5. B	9. D	13. A	17. C	21. D	25. E
2. D	6. C	10. E	14. C	18. C	22. C	
3. A	7. E	11. D	15. A	19. A	23. E	
4. B	8. C	12. D	16. B	20. C	24. B	

Section 2 – Reading

1. E	6. C	11. C	16. E	21. A	26. D	31. E	36. B
2. D	7. A	12. D	17. C	22. D	27. A	32. E	37. D
3. D	8. E	13. C	18. D	23. A	28. E	33. A	38. B
4. C	9. B	14. E	19. E	24. B	29. A	34. E	39. B
5. D	10. B	15. E	20. B	25. B	30. D	35. C	40. D

Section 3 – Verbal

1. A	9. B	17. B	25. E	33. E	41. A	49. E	57. D
2. D	10. B	18. E	26. D	34. C	42. A	50. E	58. E
3. C	11. B	19. D	27. B	35. E	43. B	51. D	59. C
4. A	12. E	20. C	28. B	36. E	44. A	52. B	60. D
5. E	13. C	21. E	29. C	37. A	45. D	53. E	
6. A	14. C	22. D	30. E	38. E	46. B	54. A	
7. E	15. E	23. C	31. D	39. E	47. C	55. D	
8. B	16. E	24. A	32. E	40. D	48. D	56. E	

Section 4 – Quantitative

1. A	5. A	9. D	13. C	17. C	21. C	25. C
2. C	6. C	10. B	14. D	18. E	22. E	
3. D	7. D	11. C	15. C	19. C	23. A	
4. A	8. E	12. D	16. E	20. B	24. B	

Section 5 – "Experimental"

1. B	3. D	5. A	7. D	9. D	11. C	13. B	15. B
2. D	4. D	6. A	8. B	10. B	12. B	14. D	16. B

Made in United States
Cleveland, OH
22 November 2024

10854326R10103